GW00362530

THE THI

TOWARDS A NUMERATE SOCIETY

THE THIRD R

TOWARDS A NUMERATE SOCIETY

Edited on behalf of the Mathematical
Education Section of the National
Association of Teachers in Further
and Higher Education

by

J. A. Glenn
Kesteven College of Education

Harper & Row, Publishers

London   New York   Hagerstown   San Francisco   Sydney

British Library Cataloguing in Publication Data

The third R.
    1. Number concept   2. Mathematical ability
    I. Glenn, John Albert   II. National Association
    of Teachers in Further and Higher Education.
    Mathematical Education Section
    513   QA141.15

ISBN 0–06–318075–8
ISBN 0–06–318076–6 Pbk

Designed by Richard Dewing, 'Millions', London
Text set in 10/12 pt Photon Baskerville, printed by photolithography,
and bound in Great Britain at The Pitman Press, Bath

Contents

Preface

This book is part of a continuing project of the Mathematical Education Section of the National Association of Teachers in Further and Higher Education (NATFHE) and is aimed at keeping discussion alive among those directly concerned with mathematical education. It is addressed to primary and secondary teachers, managers and governors of schools, and all who are concerned with numeracy in society, including its employers, its parents, and its more thoughtful citizens.

The book is an expression of collective opinion and not an academic compilation seeded with references to which many of our readers are unlikely to have access. It has not been easy to strike the intended note and to discuss essential topics without making the result read like an instruction manual for teaching arithmetic. We assume that our readers are themselves, from their own schooldays, familiar with the concepts and processes discussed, but we analyse them in some detail to show how classroom arithmetic may affect eventual numeracy. It is true that our treatment shows what sort of teaching we have in mind, but this work recommends attitudes and approaches rather than step by step procedures. It is not a programme to eliminate innumeracy, but an attempt to identify numeracy itself as it emerges from the process of education. Nor is it an attempt to limit the scope of the syllabus. Its aim is to separate the basic and essential in number skill from the wider field of mathematical education.

The text has been edited by J. A. Glenn of Kesteven College of Education, and is based on contributions by:

K. L. Gardner, Brighton Polytechnic

J. B. Hoare, Roehampton Institute of Education

J. J. Jackson, Mathematics Adviser, Norfolk

D. Paul, Trent Polytechnic, Nottingham

Hilary Shuard, Homerton College, Cambridge

W. R. Slack, Trent Polytechnic, Nottingham

The Section is grateful to the Nuffield Mathematics Continuation Fund Committee for a grant meeting the expenses of production and to Sheila Harris for her patience in typing the manuscript.

INTRODUCTION

NUMERACY AND LITERACY

Discussion of the number skills of school-leavers or of young entrants to trades and professions sets a familiar pattern. Those responsible for employing or training them complain that necessary basic knowledge is lacking. The schools are often blamed, and innumeracy may find itself associated with bad spelling, lack of discipline, or even a decline in corporal punishment. Discussion rarely includes the scope and purpose of the use of number, rarely asks how numeracy can be assessed, rarely distinguishes between computational skill and an ability to make and understand quantitative statements. The tests devised by training officers in industry may have scant bearing on any number processes actually used on the shopfloor, and it is not often acknowledged that the deficient skills of school-leavers may arise because classroom methods neither teach them in the form required nor leave the pupil with sufficient understanding to grasp alternative approaches. There is indeed a growing gap between the arithmetic of school courses and its uses in the supermarket world of prepackaged merchandise and automated check-out points. One recognizes a similar gap between the vocabularies of some reading schemes and the uses to which the child needs to put his reading skills outside the classroom.

There is, however, more to numeracy than computational skill. The ability to think in quantitative terms, to use numerical concepts with a full understanding of their scope and limits, is not widespread in society. Indeed, many who pride themselves on their speed and accuracy in totting columns of figures without the aid of calculators lack the discrimination that marks a higher order of ability: they will happily calculate a result to four figures from data accurate only to one, resisting or even resenting any suggestion that what they have produced is misleading or possibly meaningless. It is not an easy task for education to develop this second level of skill in using the language of number, although it becomes even more necessary as calculators take over the mechanics of computation.

It is also difficult to reach agreement on what needs to be taught. The examination syllabuses that guide the work of most schools, even the work of their non-examination forms, are constructed to meet the interests of specialists rather than the needs of society. Any teacher, asked to make a list of essential topics as the basis of a syllabus, will include items that few people ever use outside the classroom. For any given numercial skill, one can usually find someone who claims to need it, often an employer who bemoans the lack of it in his employees. There are and always will be needs which are specific for a given trade or industry. Much that is taught in elementary syllabuses is an essential foundation for more advanced study, even if it has no immediate application. It does not follow that such a topic is essential for a general education.

It may be that the concept of a 'general education' is meaningless. A person's general education is extended through commitment to interests that some others may not share, or the exercise of skills that some others do not possess. Nor can there be any disadvantage in knowing more about a subject than one needs in day-by-day dealings with it. For all that, one feels that it is not a waste of time to try to define basic numeracy as part of the process of education.

The word 'numeracy' is a modern coinage on the model of 'literacy', with 'innumerate' constructed to correspond to 'illiterate'. It may be instructive to make a comparison. The word 'illiterate', of course, is often used pejoratively: a badly written letter is condemned as 'an illiterate scrawl'. This is a contradiction in terms if considered pedantically. Behind it, however, is the opinion that, at least in a country where education is compulsory for all, a citizen should not only be able to read and write but should do these things reasonably well. And, although we recognize degrees of literacy, there is a fairly clear criterion for being literate, in reading if not in writing.

We need criteria for numeracy. It is easy to see that a person who cannot tell the time is socially disadvantaged, but less easy to state what number skills should be available to adults in general. By 'in general' we mean those who do not need specific mathematical abilities in their day-by-day lives, in much the same way that the general reader does not need the vocabulary of chemistry.

What schools expect of their leavers is usually based, even for pupils who never take examinations, on examination syllabuses and a long tradition of arithmetical teaching. Not only the required skills, but even the actual methods of computation, have often followed this tradition for so long that some teachers, even, imagine them to be necessary and immutable.

It is also true that schools, particularly secondary schools, find difficulty in organizing the work of their pupils so that those who develop special interests can do so without involving the majority who do not. Any solution to this problem involves not only radical changes in secondary curricula but in the assumptions on which secondary education is based. It is not likely to be acceptable unless one can show that special interests and aptitudes are not being submerged in the more general need. The whole question involves agreement on content and teaching methods, at a level of national cooperation between educators and society not called for by the problems of literacy, which are so much more clearly seen.

The constant reference to age norms underlines the difficulties. We rightly expect a ten year old to read with ease at the level of Enid Blyton's *The Secret Seven*: we should not until he was much older expect him to read, and still less to use in speaking or writing, a full adult literary vocabulary. Yet tests set for school children in arithmetic often use rates and taxes, discounts and other topics completely outside a ten year old's range of interests. Even worse, they often misuse basic numerical concepts to construct puzzle questions dubiously justified as mathematical diversions. Here is a specimen, taken from a textbook and given without comment. It deals with averages.

> During one week a man smoked an average of 23 cigarettes a day. For Sunday, Monday and Tuesday his average was 21 daily, and for Wednesday, Thursday and Friday it was 22. How many cigarettes did he smoke on Saturday?

We do nevertheless expect a numerate adult to handle civic arithmetic and to be aware of the rather limited range of numerical statements where the use of an average or mean value is permissible without loss of essential information. The very common misuse of average values in political or economic discussion is a good indicator of the nation's failure to achieve full numeracy.

This book discusses general numeracy in society as distinct from mathematical ability. As such it is more than an attempt to construct a core syllabus for schools. We are concerned not only with the child's but with the adult's concept of number, and ask what can be done in the schools to extend it. There is no question of limiting what schools attempt to do, but there does remain the problem of selecting from the language of number that part of it that should be required knowledge for an educated man. This at least needs to be taught, and taught in such a way that this section of the language of number becomes easy and familiar.

Those number concepts and processes that we regard as essential we shall discuss in depth and in detail, describing one or more of the possible lines of approach that schools have followed with success. The book is not addressed to educators as such, for whom most of the content will already be familiar, but to all those who are concerned in the quality of numerical thinking. It is certainly not a pronouncement on the way mathematics should be taught, and the many references to the classroom arise only because it is there that most of the adult's number concepts are developed.

It has been difficult to find criteria for the inclusion or omission of topics, to select from the school syllabuses what is wanted for our purpose. A useful approach is to ask, of a given topic learnt at school, whether one is in general disadvantaged by having forgotten it. Here it is the interpretation of 'in general' that takes the strain, since we have already noted that someone may be found to push the claims of each and every numerical process. It is in this way that school syllabuses inflate themselves, acquiring new items faster than their compilers are prepared to reject the obsolete. We have retained only what is basic and essential, omitting any topic that could be called marginal.

Given a sound general numeracy, any specific needs that arise should provide their own motive for learning. Most school courses are still based on the principle of 'learn now, use later', an approach that has demonstrably failed to produce sufficient school-leavers to meet even the present specific needs of industry and commerce. A possible solution to this problem, which shifts the onus for training in specific skills to the potential employers in a kind of academic apprenticeship, obviously calls for a more thorough rethinking of education, particularly secondary education, than we are currently prepared to attempt or likely to accept. This book does not argue for major changes and tries to avoid special pleading. It develops, with reference to school and society, areas of numerical skill and understanding without which a modern citizen fails in whatever is the arithmetical equivalent of speaking the Queen's English.

Many modern citizens, of course, do well enough for themselves without ever reading formal or 'literary' English. Numeracy, like literacy, forms a continuous spectrum of skill, even after one has decided on the rough stopping point at which general numeracy becomes academic or technical mathematics. At the risk of confusing schooling with education, we discuss number problems as far as we think they should be familiar to a school-leaver who has not, either as a main or a subsidiary study for examination purposes, followed a course in mathematics, the exact sciences, or any of the modern studies that develop quantitative aspects, and has thus not been faced with other than general requirements.

A good primary pupil will probably have learnt most of the necessary manipulative skills by the age of eleven and will gain in numeracy as he matures. Some children are later in starting or slower in developing and will need more time. The secondary schools will have to take them over and arrange programmes to complete the process.

CHAPTER 1

THE LANGUAGE OF NUMBER

The words and the symbols

We need to discuss the very early stages of learning about number to see
where the difficulties begin. As a step in a child's mental development,
simple number operations symbolically expressed, such as 'sums' of the
form $2 + 3 = 5$, often seem to cause less difficulty than the early stages in
reading. Indeed, a less able child may acquire an early skill in writing out
these number statements that eventually acts as a block to achieving true
numeracy. Finding reading difficult, he cannot cope with problems that use
numbers in a verbal context, and may be allowed to complete page after
page of sums when he should be giving more time to reading. Because he
may never learn to relate the symbols effectively to situations involving
number, he does not develop a sense of number. He may lack, for example,
the judgement that rejects as obviously wrong the result of a large error in
calculation. Where infant and junior schools can work together, with the
one carrying on smoothly the programme begun in the other, it is possible to
delay the teaching and drilling of formal arithmetic until reading skills are
better established. The younger children learn the number symbols as convenient abbreviations for the words, learn to use them in recording, and can
complete, informally and with the aid of apparatus, such statements as
$2 + 3 = 5$. The apparatus may be boxes of shells and conkers, or the commer-

cially produced 'structured material' of rods or cubes. In this way, the child does begin to learn, in context, the first number bonds, but not in the first place as abstractions. The route is from a sentence such as

These two conkers and these three conkers make five conkers

through the statement

Two and three make five

to the symbolism of

$$2 + 3 = 5$$

The child is not trained to remember the symbolic statements by completing page after page of examples but learns to use the symbols as a result of counting.

The distinction is fundamental for numeracy. Unfortunately the routine training is quicker and gives more immediate satisfaction to those concerned. A criticism often made by schools against schemes for number work is that the children can do the arithmetic but cannot read the text or the problems. If at this stage the teacher demonstrates the arithmetic and allows the pupil to skip the text, then he is beginning to open the gap between literacy and numeracy that a balanced education tries to avoid. Given that the text is carefully written using the simplest vocabulary and constructions that will allow its concepts to be expressed, the pupils' first books of number work are also among their first books of language in use. The prose that expresses the number work is not part of a reading scheme of simple words devised to be read by a phonetic or any other method, but an attempt to use language to convey instructions and information. If the children cannot read them, their reading is not, we maintain, sufficiently far progressed to do meaningful arithmetic. Indeed, a well-written arithmetic text could be part of the reading scheme, so that the use of the skill goes along with its acquisition. Innumeracy, like many other defects of education, probably owes much to the subject divisions still imposed on the learning process, and to the temptation thus put on schools to go ahead along lines that offer immediate success, rather than to make a planned advance on all fronts.

Counting

Number words emerge in language to serve the separate but related processes of counting and ordering, and make precise for the needs of a developing society the vague concept of plurality. That the process of counting became fully systematic at a late stage in the development of language is shown by the irregularities in the pattern of the number words compared with that of the symbolic notation, and the many words that refer back to counting in braces, dozens, scores, and the like. It is odd that the classical languages show less evidence of irregularities than modern English or French, but this could well be because they have come down to us in literary rather than colloquial forms.

Most of us can count with ease. There is no difficulty and the whole process appears open and unremarkable. Very young children often fail to count properly and it is instructive to enquire why their totals do not agree with ours. Sooner or later they realize what is required of the counting operation, and it is doubtful whether attempts to get them to perform successfully earlier in life are effective. Nor does it matter whether we believe that the numbers are necessary and immutable, independent of the physical universe, or whether we think of them as contingent. There is a delightful argument in *Tristram Shandy* on this issue. The modern view considers number as a linguistic convention that works partly because its structure models the given structure of the actual world in which we live, and partly because we have all come to agree on exactly the same convention. The basic fact is that we do all count in the same way.

We feel, nevertheless, that the teacher at least needs to consider the counting process. The importance, and indeed part of the justification, for teaching about the formal concept of *set* in mathematics, is that it gives rise to the language in which counting and number can be discussed. A set is a collection of objects of any kind, real or abstract, brought together for consideration as a single entity. The simple properties of sets and the way of combining them together give rise to what is usually called set algebra.

As an easy topic in which children can make rapid progress, too much time is sometimes given to work in set algebra at a primary level: it has become one of the 'gimmick' topics of modern mathematics. Because set theory is important in the foundations of mathematics, it does not follow that it is important in school. There is, however, one set operation that eventually leads to the number concept.

Given a set, its members may be matched up one at a time with those of another, as a set of people may be matched up with a set of chairs by asking them to sit down. If everyone is seated and there are no spare chairs, we can then say that the two sets 'have the same number of members'.

We do not need to count to establish this fact. It is not necessary to know what the actual number is. The next step is to take a standard set of words, called number words, always taken in the same order and learnt by heart. If we now match up the set of chairs or the set of people with the set of words, by pointing or touching as we say them, sooner or later we come to the last member of the set. This is always possible since the list of number words is constructed so that it can be continued indefinitely. Then the number word we have reached when we point to the last item is the number of the set. The set of words, of course, is the sequence

one, two, three, four . . .

We label the set of people or chairs with the number word that matches up with the last one counted. It is in this sense that we speak of a group of 'seven' people.

The process of thus matching two sets of the same count is called one-to-one correspondence: it is a very important concept in mathematics, but can be taken as quite straightforward in ordinary speech. Once described, it makes clear exactly why tiny children count erratically. They either miss out or repeat number words from the recitation, or they miss or count more than once some of the items. We ourselves find it difficult to count moving objects such as birds in flight. Most children learn the number words correctly before they can establish techniques for successful counting.

The word 'technique' seems pretentious here, but only if we think of counting as dealing with small convenient groups. If we ask half a dozen people to count, independently, arrivals for a large concert or the items in a small factory stockroom, we soon see that a technique is necessary. There will probably be six different counts, and it is a first lesson in numeracy to realize that only one can be right, although all may be, and probably are, wrong. We strongly recommend that children should be taught to collect statistical information of this kind, using tallies, or hand tally counters, recording the information in tables and graphs. Where practicable, more than one pupil or group should repeat each count, not only to show the importance of checking but to impress on them the existence of errors.

Sorting

Sorting is an essential prerequisite of counting. One cannot eliminate sorting. Whether one touches things in order or merely looks at them one by one, or even listens to them as in counting the strokes of a bell chime, one is always distinguishing the things counted from those ignored.

It is not just a matter of rejecting the unwanted. How one accepts what one accepts is part of the process. How can one count the objects in a child's toy-box? There is a toy car: does one count the wheels? Is the question settled by the fact that the wheels do not come off, so that the car is an inseparable whole? But the tyres do come off. Does this make four more items in the count? Is the pack of playing cards in the box one item or fifty-two, plus a carton and two jokers? Does one count the fluff in the corners? The Greeks, who began mathematics as we know it, did not consider these questions trivial, and nor should we. They do not have self-evident answers: the person counting has to make a decision. What counts when one is counting?

It is not suggested that a child should postpone counting until some specified level of sorting is achieved. Counting in general will not let itself be postponed. It is also clear that sorting or classification is fundamental to language: the difference between the specific and the general in naming objects, in distinguishing dandelions from daisies, and in deciding whether to call both flowers or weeds. Children, we suggest, should have extended practice in sorting and classifying outside the context of counting. Activities which require sorting should initially be undertaken without the additional burden of counting. There are many sorting games known to teachers, and purpose-built apparatus is on the market – logic blocks and the like – which simplifies the task by restricting choice to a few chosen attributes such as colour, thickness, shape, or size.

For the purposes of counting, sorting needs to be further considered and refined. Usually we need to be able to make an unambiguous decision. If we ask children to put their hands up if they have a dog, we should get the result intended. If we ask them to raise their hands if they have a dog and keep them down if they have a cat, those with both or neither will not know what to do. In terms of the language of sets, we need to make a 'complementary partition' of the set of children, one which splits them into two groups without overlap. It is not, however, essential to use this technical language in the classroom, and indeed its unmotivated introduction is one of the many aspects of modern mathematics teaching open to criticism.

Ordering

The number words have two distinct aspects, suggested by the two words *cardinal* and *ordinal*. The 'cardinal number' is the number label attached to any set on counting, as already discussed. The word ordinal implies order. Counting uses the ordinal sequence of the numbers to arrive at the cardinal of the set of items counted. We also use the standard order of the number words to sequence the members of a group or collection, often using the special ordinal forms *first, second, third, fourth* ... In a suitable context the two forms are equivalent. For a row of soldiers already numbered off the two orders

Fall out the third man

Fall out number three

produce the same result.

Certainly the conventions under which these words are used should be discussed, particularly among older pupils and students of language. A diagram showing ordinal six and cardinal six will illustrate the two uses.

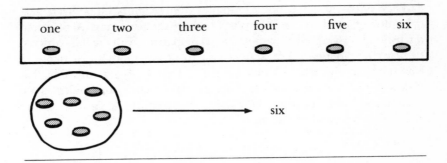

The difference was no doubt important to the members of a Roman cohort on parade and about to undergo decimation, the military punishment that executed every tenth man.

It is also easy to show how important order is to a ordinal sequence. Sets of dots or counters to illustrate the cardinal numbers can be put in any order.

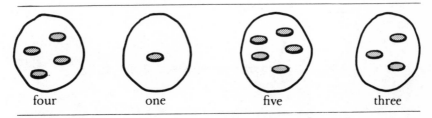

One cannot rearrange the number sequence without making it meaningless for purposes of counting.

five   two   six   one   four

The considerations are important for a mature understanding of number and for the teacher who has to guide the young child towards it. Children can often be seen confusing ordinal and cardinal thinking. A six year old, holding up the last of a set of six articles she was told to count, asked why she had to say 'six' when it was only one. The question would have delighted Socrates.

Children's experience of ordering needs to be as wide as possible. Numerical order then takes its place in the extended mathematical and practical concept of an ordering relation. Attributes such as age, size, loudness, mass are readily available. Attributes where discrimination is difficult provide useful subjects for discussion and eventual agreement or disagreement. Examples are fairness of hair colour, pleasantness of taste, darkness of colour. Some of the examples given might result only in a partial ordering, as when two objects in a set have the same mass. Activity and discussion involving such examples, and comparison with the fully ordered sequence of number words, should be part of everybody's experience in education. It is practice in that particular kind of logical thinking that makes full numeracy possible. As a matter of classroom practice, we recommend from an early age the use of the symbols > (greater than) and < (less than) in number statements.

$$9 - 3 < 8$$

in context could be as important a statement as

$$9 - 3 = 6$$

Its use certainly offers extended practice in writing and using number facts, and it enables a set of any numbers to be put into an ordinal sequence, in which each number is less than its successor.

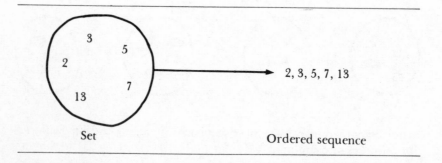

Action for children: a summary

The action to be taken, either in school or later, can best be summarized by a flow diagram. It differs from many flow diagrams in that some of the arrows are drawn in both directions, to show interaction rather than chronological sequence. Here is the diagram, with the school programme below the dotted line:

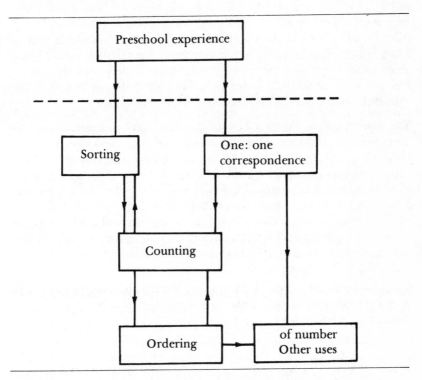

Although many of the finer points of the broad categories in the boxes must await mature thinking before they are fully absorbed, the diagram shows the necessary background for numeracy. Until a child can show by his actions that he has grasped these concepts, he is not ready to begin formal arithmetic.

Inevitably, and most desirably, he will begin to acquire number facts as a result of counting and combining sets of objects, but this is not the same as a premature attempt to learn tables by heart. He will also learn to recognize at once, without counting, irregular groups of two or three objects, or larger groups of six or more if, but only if, they are arranged in recognizable patterns. The impact of the Gestalt account of perception has been important for teaching at this point, even if it has been a matter of dispute in psychology itself.

The flow diagram links the prenumerate stages of number skill, in which the numbers appear as words in a verbal context represented by the number symbols only for convenience. Every normal child sooner or later will be able to carry out the required operations of sorting, counting, ordering. We do, nevertheless, believe that a deeper understanding of these three fundamental processes is involved in being fully numerate, just as being fully literate requires more insight into language than merely being able to read. Any classroom activity that furthers this deeper understanding is worth while, although it is not an end in itself and must lead on to the extended number skills, that make up the art of arithmetic.

The importance of the first box in the flow diagram cannot easily be overestimated. The knowledge and skills already possessed by a five-year-old child whose home circumstances do not leave him educationally or socially disadvantaged are considerable. Before they begin school, some children may be well advanced and even adept at the three basic processes of sorting, counting, and ordering. We also believe that most children would be better served if their parents gave them plenty of activity material collected (or perhaps devised) for sorting, counting, and ordering, rather than attempted to make them beat the school starting gun by training in arithmetic before the age of five.

We wish to stress the importance of sound numerical knowledge and adequate computation skill, and yet dissociate ourselves from a too early attempt to achieve it. The seeds, not only of later innumeracy but dislike of mathematical thinking generally, are sometimes sown by the anxiety of primary teachers to meet the external imposed criteria of selection tests. There are signs that parental or even political pressures may replace the pressure of the tests.

The reader will see that the diagram has a sixth box labelled 'other uses of number'. Because the number words and their symbols form a convenient standard sequence which can be continued indefinitely, they are often used to mark or otherwise label both sequentially and uniquely. Hence members of large organizations have works or service numbers which cannot, like names, be duplicated. The numbers can be arranged in the records in an ordered sequence, and any given number can, if in place, be found by anyone who knows the number system. The letters of the alphabet can be used in a similar way, but only for smaller collections.

Such usages are not usually difficult even for those not fully numerate, but we mention them as an application of the unique properties of number that ought not be passed over.

CHAPTER 2

FORMAL ARITHMETIC AND THE NUMBER BONDS

Arithmetic as language

There are three stages in the evolution of arithmetic relevant to social numeracy, and these have already been mentioned. The first is the use of number words within the ordinary context of speech or writing, as when one says of a couple that they have four children, two boys and two girls. The second is the abstraction of the number concept from such statements freed of dependence on actual objects counted. This stage is tacitly accepted by most people, and only causes a passing academic difficulty to some who want to argue that one cannot possibly have 'two': it must be two of something. The third stage is the replacing of the verbal by the symbolic, as in

$$2 + 3 = 5$$
$$2 \times 5 = 10$$

At this stage true arithmetic begins. With the familiarity of use, the symbols begin to take on a life and a structure of their own. Most older children and adults complete a statement such as

$$\begin{array}{r} 2 \\ +3 \\ \hline \end{array}$$

by writing in the sum 5 without verbalizing the number symbols as number words.

But the fact remains that they *are* number words, and arithmetical expressions containing them can be set out, often of course with extreme wordiness, in ordinary prose. To do arithmetic efficiently one must be able to complete numerical statements without passing through equivalent literal forms, but to do it with understanding one must realize that the statements exist as language and be aware of their syntax.

The art of arithmetical education lies in the progression through the three stages outlined, in devising a programme that lets children make the transition from words to arithmetical operations. As a typical learning activity, we can refer to the use of number boxes, each marked with the word and the number symbol, which children fill each day with suitable objects until the association of counting word and symbol is firmly made. We recommend as essential a box which is always left empty, marked with the two words 'none' and 'zero' and the symbol 0. The two words are by no means equivalent, but both need to be associated by the child with the symbol. Many counting schemes using pictures or jingles are deficient in omitting the null count, and most fail also in not grouping unlike objects to illustrate number words. This is quite natural. Most of us actually see the pattern as three noughts and two crosses rather than as a group of five

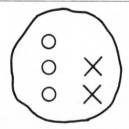

marks. Nevertheless, a miscellaneous group can have the same cardinal number as a group of similar objects, and it should be part of the child's developing concept of number to understand this. Once the number words and symbols are known, at least to a count of six and more practically to a count of ten, investigation of number facts can begin. At this stage the symbol 10 is presented as a compound symbol made up of two separate marks but standing for a single concept, like the Chinese ideogram for 'button'.

We do not discuss the place value notation that separates the digits into columns. The symbols from eleven to twenty can be thought of in the same way. Obviously the child is likely to see that there is a repetitive pattern here and may talk about it. The point, however, is that we do not teach it: we do not say at its first introduction that ten is written as the symbol for one followed by zero.

The number facts

The number facts arise from the counting process applied to more than one set of objects. Whether the child begins to investigate them before he is completely familiar with numbering and ordering up to ten, or whether he delays it until he can count and record unerringly up to twenty, is a matter more for the teacher's own judgement than for debate. The work can begin when the child is completely at home with numbers up to six, extending its range as each new number symbol is firmly established. What matters is that the number facts should be kept within the limited set of those known and familiar to the child.

By a 'known' number fact we mean a number statement *that is known and learnt by heart to the point of instant recall*. These are the essential and irreplaceable starting points of all computation, and without them speedy computation is impossible. Their learning is a critical point in the acquisition of numeracy, and calls for all a teacher's professional skill in timing the work programme. Pushed ahead too quickly, the number facts outrun the empirical experiences on which they are based. Left too late, not only do they become harder to remember, but the pupil acquires techniques for recovering the facts without instant recall. These techniques, such as 'counting on' give the correct result and are difficult to detect in operation, but they are slow and inefficient when the facts are needed for the purposes of computation.

In the light of the arguments about innumerate school-leavers and 'teaching them their tables', it is often forgotten that all the basic number facts, the ones that must be known by heart before efficient computation can begin, can be written on a postcard. We shall give them as double entry tables before discussing what can be done to see that all children know them.

second number

+	0	1	2	3	4	5	6	7	8	9
0	0	1	2	3	4	5	6	7	8	9
1		2	3	4	5	6	7	8	9	10
2			4	5	6	7	8	9	10	11
3				6	7	8	9	10	11	12
4					8	9	10	11	12	13
5						10	11	12	13	14
6							12	13	14	15
7								14	15	16
8									16	17
9										18

first number (rows)

The 55 addition facts (or number bonds)

second number

×	1	2	3	4	5	6	7	8	9
1	1	2	3	4	5	6	7	8	9
2		4	6	8	10	12	14	16	18
3			9	12	15	18	21	24	27
4				16	20	24	28	32	36
5					25	30	35	40	45
6						36	42	48	54
7							49	56	63
8								64	72
9									81

first number (rows)

The 45 multiplication facts (or products)

This gives one hundred facts in all, involving the addition or multiplication of pairs of single-digit numbers. Twenty of these, the results of adding zero and multiplying by unity, do not change the operand (i.e., the number being multiplied or added to). Ten others appear in both tables, since a result such as $3 + 3 = 6$ is equivalent to $2 \times 3 = 6$. The seventy that are left must be taught and learnt, *and there is no possibility of compromise.*

Any discussion of the teaching of arithmetic must begin by recognizing that these are, psychologically, 'atomic' facts. They can, it is true, be derived as needed by counting and combining sets of objects, or by a counting on process with numerals. In this way, one can 'calculate' these facts, in the original sense of the word calculate, which was to manipulate a pile of pebbles (Latin: *calculus*, a small stone) used as counters; but in the actual practice of arithmetic one begins with the known facts, and any process of calculation would be intolerably tedious if one did not know them. It is in this sense that we call them 'atomic'. The point must be made and kept quite clear. A child who makes dots or touches his fingers to arrive at

$$5 + 4 = 9$$

does not know this number fact, but merely has a mentally inefficient way of dealing with it. On the other hand, nobody is likely to know the product 2967×429, although most of us know a method for obtaining it. One excepts the young lady in Ionesco's play whose memory contained such information because, not being able to do arithmetic, she had learnt it. As she herself said: '. . . I have learnt by heart all the possible results of all possible multiplications.'

It follows that, since learning the number facts must come before arithmetic proper, we must see to it that all children acquire them. This does not mean that arithmetical studies cannot begin until the seventy facts are mastered. It does mean, however, that care must be taken not to allow work in formal arithmetic to outrun the build up of an efficient memory-store of number bonds. As a matter of classroom strategy, one may want to keep it just ahead to provide both motive and practice, but this should be a planned and controlled exercise in pacing.

In what follows we use technical words such as inverse, complementary, commutative. Teachers will of course be familiar with them, but we hope that the general reader will find them explained in context. These are not to be used explicitly in the primary classrooms, but children should have an implicit understanding of them.

Addition of single-digit numbers

Given that the children know the number words and symbols to an agreed minimum as discussed on page 18, they can begin to learn the addition facts by combining counted groups of objects, recounting, and recording, all within the planned limits. Apart from counters, shells, conkers, and the like, there is a wide range of commercially designed apparatus available. Because children usually enjoy using this material, activity is sometimes dismissed as 'playing around instead of working'. The apparatus is, of course, intended to be part of a planned programme of learning controlled by the teacher.

Children should use all three usual forms of recording the results

1. $4 + 5 = 9$

2. $(4, 5) \xrightarrow{\ +\ } 9$

3. $\begin{array}{r} 4 \\ +5 \\ \hline 9 \\ \hline \end{array}$

The second form may be unfamiliar to the reader. It brackets a pair of numbers together, and associates with this pair the number that is their sum. The symbol $\xrightarrow{\ +\ }$ is read as 'added together give'.

The children will only begin to remember the bonds after they have had plenty of practice in recording the results of their counting, but if the work is organized properly, it is self-progressive: the child who knows the sum of 4 and 5 will write the result immediately.

It has never been considered necessary to make children learn these bonds in table form, although this has certainly happened with the multiplication bonds, one of the notorious 'sticking points' in becoming numerate.

The task of memorizing can be made easier by using, once the bonds begin to be known, a systematic record.

Making up 2

$1 + 1 = 2$

Making up 3

$1 + 2 = 3$

$2 + 1 = 3$

Making up 4

$1 + 3 = 4$

$2 + 2 = 4$

$3 + 1 = 4$

Making up 5

$1 + 4 = 5$

$2 + 3 = 5$

$3 + 2 = 5$

$4 + 1 = 5$ and so on.

We stress the need for the child to begin by using real articles. The development should be slow and practical. Although the child may be able to recognize the numbers up to the two-part symbol 20, we only consider sum of pairs up to (9, 9) since (10, 10) is to be tackled later when place value is being studied. One would also omit, at this early stage, pairs such as (0, 2) or (2, 0) since they would appear pointless to the child. There are thus 81 bonds to learn, but once the child has realized the commutative nature of addition, that is, that the order of the pair can be reversed without changing the sum, as, for example

$$4 + 5 = 5 + 4$$

there are only 45. The child should eventually be competent to give these verbally and in writing without help of apparatus. If he runs into any difficulty, he should be referred back to trial and counting, and not merely be told and made to repeat.

The child will also deal with the sum of three numbers, and learn that the bonds are always dealt with in pairs.

$$2 + 3 + 4 = 9$$

is done in the two mental stages

$$2 + 3 = 5$$
$$5 + 4 = 9$$

One could also say

$$3 + 4 = 7$$
$$2 + 7 = 9$$

This corresponds, in mathematical language, to the associative (or bracketing) rule for addition, which tells us that

$$(2 + 3) + 4 = 2 + (3 + 4)$$

The term should not, of course, be used with children, but they should have plenty of number activities in which the rule is implicit. It is important because some arithmetical processes, such as subtraction or division, are *not* associative.

First stages in multiplication

From this point some teachers like to consider subtraction, but we prefer to follow with the simpler concept of multiplication as repeated addition. Children will have such results as

$$3 + 3 = 6$$
$$4 + 4 = 8$$

Expressed in terms of the objects counted, these become

two lots of three are six

two lots of four are eight

The children will then get, as well as two lots of a given collection and its symbolic record as above, such number 'stories' as

$$1 + 1 + 1 + 1 + 1 + 1 = 6$$
$$2 + 2 + 2 = 6$$

These are now verbalized as 'six lots of one', 'three lots of two', and so on, and the children can then pass to the symbolic forms.

$$6 \times 1 = 6$$
$$3 \times 2 = 6$$
$$2 \times 3 = 6$$

and so on. These are the products or multiplication facts.

We recommend that the operators $2 \times$, $3 \times$ ... are read 'two lots of ...', 'three lots of ...', since this relates to the original situations from which the results arise. Later, these forms can be contracted, giving such forms as 'two sixes', 'three sevens'. The results can be referred to as *the product of* the pair, but the technical word is best delayed till most of the facts are mastered. What must be avoided is the scarcely literate 'times', a coinage of hard-working but not always well-informed Victorian schoolmistresses passed on from generation to generation. One still hears a complete list of the products of three called a 'three times table'. Such a list, which is certainly useful for reference, is best called a 'table of threes', which describes it without doing violence to our syntax.

Once again, children will discover the so-called commutative rule for multiplication as given for addition on page 22, and will soon be able to write both

3 × 4 = 12, and

4 × 3 = 12

knowing that each follows from the other, but without ever meeting the word 'commutative'.

We strongly recommend beginning this work using no products larger than 18, the maximum for the additive bonds. In this way, we have as it were bitten off for chewing the easier 22 of the 45 multiplicative bonds. Of these, 9 are the products 1 × 3 = 3, 1 × 6 = 6, and so on. There are only 13 facts presented for memorizing.

		first number								
	×	1	2	3	4	5	6	7	8	9
	1	1	2	3	4	5	6	7	8	9
second	2		4	6	8	10	12	14	16	18
number	3			9	12	15	18			
	4				16					

Reference table of beginner's products

Several sets of well-designed structured apparatus are available to help learning with controlled activities and the children should eventually learn these bonds. They should also be able to recall or write them in any order, random or systematic. It is this skill that underpins later work in computation.

From this point we recommend an extension of activity with additive bonds to cope with subtraction, rather than continued learning of products. This work can well begin before the last of the 22 products is securely remembered.

Aspects of subtraction

Although adding and subtracting are complementary processes in arithmetic, the conceptual stages of the latter are much more difficult, and there is no single physical activity corresponding to them.

Because of its relation to addition, subtraction does not introduce any new number facts: it is merely an alternative expression of known facts. Because of the physical ambiguity introduced, children should work out the subtraction facts very slowly and thoroughly. We suggested a break after the addition facts, with an incursion into multiplication, in part to allow the children to gain greater confidence in handling numbers before tackling the difficulties of subtraction.

The problem is easily seen. Suppose one has two sets of counters A and B as in the diagram, to show counts of 5 and 3:

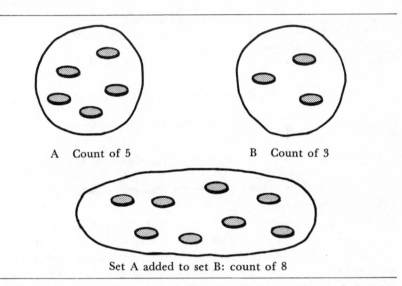

A   Count of 5   B   Count of 3

Set A added to set B: count of 8

Set A added to set B gives a count of 8. The diagrams make an obvious interpretation of the instruction 'add set B to set A'. But how can set B be taken from A? They are already separate. If the diagram satisfactorily represents 5 + 3 = 8, what diagram shows the difference 5 − 3? If the reader finds that the following discussion calls for close attention, it will help make the point that subtraction is a more complex process than addition.

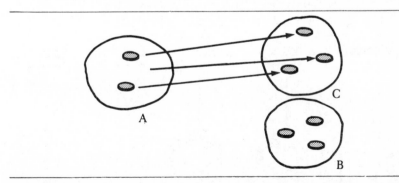

Consider as before set A of 5 objects and B of 3. One can remove from set A a set C having the same count as B, that is, the members of C can be put into one-to-one correspondence with those of B. The set A now contains only the members shown, and counting or recognition gives

$5 - 3 = 2$

It probably helps to arrange the counters in rows

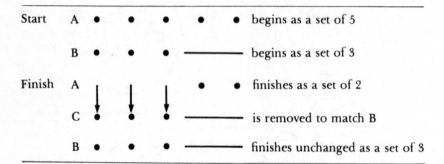

Start	A	• • • • •	begins as a set of 5
	B	• • • ————	begins as a set of 3
Finish	A	• •	finishes as a set of 2
	C	• • • ————	is removed to match B
	B	• • • ————	finishes unchanged as a set of 3

The discussion is exactly as before. We rarely have occasion to use subtraction in this way with a physical set B of the same kind of objects as in A. If one has a set of 5 and is told to 'take away 3', one does *not* normally use a representation of the second set. One counts up to 3, numbering off each member of the first set as they are moved. This process requires the child to store the second number in his head during the operation, and one can see that it is already at a higher level of abstraction than addition. The commonest practical situation involves two different sets.

Example: There are 5 fruit salads left at the
serving counter but 3 children have not
yet taken theirs. How many salads will be
left when all the children are all served?

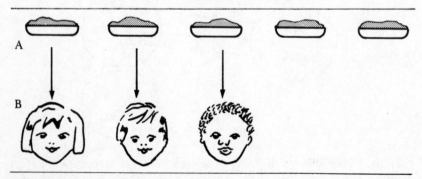

as before

$5 - 3 = 2$

This time A and B are sets of different objects – children and fruit salads. There is to be a one-to-one correspondence between the members of B and *part* of the set A, and what concerns us is the numerical difference between them.

It is worth noting that in practical addition the two sets are usually of similar objects. We might add a known number of boys to a number of girls to get a total of children, but we would not add children and fruit salads. In most of the actual physical applications of subtraction we are either 'taking away' using a count or following a procedure as in the last example where the two sets are different – physically as well as numerically. At later stages both addition and subtraction are used not with actual objects but with abstract units of measure, as in finding a total length or the difference between two lengths.

There is, however, another representation which is important because it corresponds to a different mental approach to the number bonding.

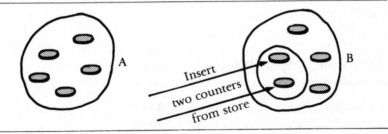

Here we put in extra counters from a spare pile until B has the same count as A. The added counters form the set C, shown included in B. One sees that we have arrived at the result

$5 - 3 = 2$

indirectly by saying

$3 + 2 = 5$

The operation gives the difference between A and B as before. Discussion of these physical representations tends to be more confusing for an adult than the abstract process with number that has become familiar to him through repetition. The multiple interpretation, however, shows that the subtraction concept, if analysed, is indeed more difficult than addition. The final test of a person's understanding comes later, but only if he continues his mathematical studies and needs to work with negative numbers. Moreover, in the mental equivalent of the last representation, to complete the statement

$$\begin{array}{r} 9 \\ -4 \\ \hline \end{array}$$

one does not say, mentally, 'nine take away four', 'nine less four', or 'four from nine', but 'four *up to* nine'. One writes down as the answer the number which must be *added* to four to make it up to nine. For this reason we say we have subtracted by *complementary addition*.

We recommend that children should work with both aspects of subtraction, the 'taking away' and the 'building up', acquiring the numerical skills slowly but steadily with constant reference to concrete operations.

As each subtraction fact is investigated it is seen to pair up with an addition fact. The subtraction

$$12 - 7 = 5$$

implies that

$$5 + 7 = 12$$

So also does the subtraction

$$12 - 5 = 7$$

In fact each addition statement corresponds to two separate subtraction statements. Given a set of three suitably chosen numbers, there are four statements that can be written, because we also have

$$7 + 5 = 12$$

When children begin to work on the subtraction facts, they should keep addition in constant revision, and should often be asked to write down all four statements. Thus, for the number triple (4, 5, 9) they will write, as four variants of the one fact:

$$4 + 5 = 9$$
$$5 + 4 = 9$$
$$5 = 9 - 4$$
$$4 = 9 - 5$$

Note that adding five to four gives nine: subtracting five from nine gets us

back to four again. In mathematics a second process that brings us back to
where we were before we applied the first is called the *inverse* of the first.
The child thus learns subtraction facts as inverses of the corresponding ad-
ditions, although he does not use this term.

Since the child should already be quite familiar with sums up to 18,
differences can be in the same range: it is here that the mental process of
complementary addition comes into its own. Suppose we have

15
−8
——

If the 45 number bonds are known as they should be, the child knows by
recall that $8 + 7 = 15$, so that 'eight up to fifteen' is seven, and in goes the
answer. It should cause no more difficulty than

$8 - 5 = 3$

given that the child comes to such results only when he is quite familiar
with the one-digit examples. The unfortunate adult who says '8 from 5
won't go, borrow 1' has had a deprived childhood! We have still not
finished with subtraction. If they work properly with counters, coloured
rods, or structured apparatus, children will soon realize that subtraction,
unlike addition, is not commutative.

$4 + 3 = 3 + 4$
$4 - 3 \neq 3 - 4$

In fact, we cannot take four from three without introducing an entirely new
set of concepts into the number system. Such new concepts we do not
regard as part of general social numeracy. The statement 'four from three
is minus one' is meaningless jargon, unless set carefully into a context of
higher level activity. This can in fact be done in primary schools among
able children, but it is not something we must make every effort to teach to
all children.

We suggested that the symbol 0, called zero or nought, should be taught as
the cardinal corresponding to 'none' or 'empty'. Note that we can say we
have three sweets but not that we have zero sweets: 0 is the only number
symbol with its own name, a circumstance that would please Humpty
Dumpty. It is in subtraction that children first encounter the need for zero
in arithmetic. The difference between the counts of two sets of the same
count is zero.

$$6 - 6 = 0$$
$$3 - 3 = 0$$

Their work should include frequent examples of this. From this point we recommend that children should have constant practice in building up given numbers in as many ways as possible within the limits already reached. They can be made to write all the important bonds by giving them the required uncompleted statements using the now well-known box notation. The child has to fill in the box.

Thus

$2 + 4 = \square$

$6 - 3 = \square$

$1 + \square = 6$

$6 - \square = 4$

$2 \times 3 = \square$

$\square \times 2 = 6$

We make no apology for repeating that if the child has any difficulties he should be referred to actual objects like shells, conkers, or purpose-made apparatus, and not merely be told what to write.

Aspects of division

We ask the reader to gather together six similar objects and try to use them as a demonstration of the number fact $6 \div 2 = 3$. He will see that there are two distinct processes, which can be shown by examples.

Example 1:   How many children can be given two apples each from a bag containing six?
The diagramatic representation is

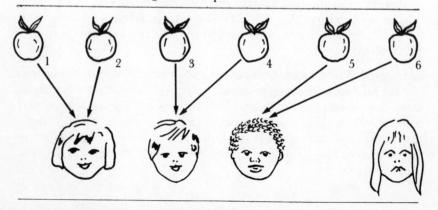

Here we give each child in turn two apples until the stock runs out. Three children get them and we can write

$6 \div 2 = 3$

We are, in fact, subtracting 2 repeatedly until none is left, and this aspect of division is usually called repeated subtraction; it may be compared with multiplication as repeated addition.

The second aspect is rather different.

Example 2:   Six apples have to be shared fairly among two children.
How many does each get?

Here is the diagram

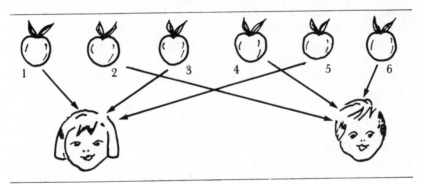

We give each child one each in turn until the stock runs out, but we still write

$6 \div 2 = 3$

Children should work on the division facts using both these processes. There can be no doubt that the physical situations are different. In one, three children get two apples each; in the other, two children get three apples each. The numerical answer stands for children in one case and apples in the other. One sees that the first corresponds to three lots of two, the other two lots of three. That is we have both

$3 \times 2 = 6$

and

$2 \times 3 = 6$

We have here a new physical illustration of the commutative property of multiplication. As with addition and subtraction, multiplication and division are inverse processes of which only one is commutative. Given a suitable triple such as (3, 4, 12) we have the four statements or number bonds, as with the first two processes.

$$3 \times 4 = 12$$
$$4 \times 3 = 12$$
$$4 = 12 \div 3$$
$$3 = 12 \div 4$$

Once again it is clear that $12 \div 3 \neq 3 \div 12$

This time the status of $3 \div 12$ in the number system cannot be brushed aside as irrelevant to general social numeracy as can $3 - 4$, but it should be obvious that we are not dealing with a cardinal or counting number, arrived at by putting a set of objects into correspondence with an ordered set of counting words called numbers. The attempt to divide any number by a larger number than itself gives rise to the concept *fraction*. This is both mathematically important and practically useful, but it cannot be introduced incidentally and without careful progression. It is dealt with in Chapter 5.

There is a further outcome of the division process, whether seen as continued subtraction or division into equal shares. If we have eleven apples, we cannot give two each to a group of children nor share them out equally among three. There is always an apple or apples left over. Any two numbers whatever may be multiplied, but either the division cannot be done at all if the dividend is the smaller, or there may be a remainder. This situation does not arise if the only quotients recorded are inverse to previously determined products as recommended.

Summary

A summary is called for at this stage, not so much to recapitulate what has been done as to make clear the upper limits of what should have been attempted. All the work suggested in this chapter has been directed towards the number bonds considered as established facts and learnt by heart. No arithmetical manipulative methods have been introduced, although all the four processes that the child will later study have been worked at in some detail. In particular, products and quotients (the results, that is, of multiplying and dividing) have been restricted to those which can be handled directly, not involving numbers greater than those produced by the additive bonds. The work has been arranged to bring out the underlying structure of the number system in its simplest form, the natural numbers used as cardinals in counting or as a standard sequence for establishing ordinality. Using our useful technical terms, we can say that the two noncommutative processes, subtraction and division, appear as inverses of the basic processes of addition and multiplication. Multiplication is the result of repeated addition.

It will be seen that we recommend an extensive programme of learning and activity, and insist that children should embark on this and complete it. Until they have done so, they are not ready for arithmetic proper, and any attempt to force them into it will sow the seeds of the numerical difficulties and ineptitudes that exist only too obviously among young and old. The criterion of success can be quite objective.

The child should be able to give, orally, any one of the 55 additive bonds or their inverses, or any one of the 22 products and quotient inverses, and should be able to complete any written exercise that requires this number knowledge, provided that the exercise is so framed that the child knows what is required.

Examples of the many ways in which questions could be framed are:

1   What is three and four?

2   Take five from twelve.

3   What added to fifteen makes eighteen?

4   What are three lots of four?

5   How much bigger is eleven than nine?

6   What is eight divided by four?

Examples of the wide range of possible written work, with verbal instructions if necessary, would be like those which follow:

1   Complete these statements by a number in the box:

   1   $6 \div \square = 3$ 4 $7 \times 2 = \square$

   2   $2 \times \square = 6$ 5 $\square \div 2 = 4$

   3   $6 \div \square = 2$ 6 $17 - 4 = \square$

2   Complete these

   1   2   2   12
        $+5$ -3
        ―――   ―――

$$3 \quad \begin{array}{r} 13 \\ +4 \\ \hline \\ \hline \end{array} \qquad 4 \quad \begin{array}{r} 9 \\ -9 \\ \hline \\ \hline \end{array}$$

3   Link each number in set A to a number in set B by an arrow which reads 'add one'

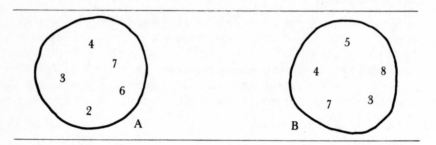

4   Draw an arrow from each of these numbers, so that it means 'three lots of', and fill in the number at the end of it. One of them has been done to show what you have to do.

1

2 ─────────► 6

3

4

5

We stress that verbal instructions should generally be used. The tasks themselves are simple and test the required skills, but not all children at this stage will be able to read the instructions. The reader will note that we are aiming at recall of each fact as it is required. This is not the same as table learning which only tackles facts in some preestablished order.

A flow diagram will show the interrelations of the topics so far considered in their precomputational stages. The work of Chapter 1 is above the dotted line.

The horizontal links are important and show the topics as inverses of one another. Readers will see that reversing the arrows in the written exercises above then calls for the inverse process to be done. Any further discussion of inverse processes is not part of general numeracy but of mathematics proper.

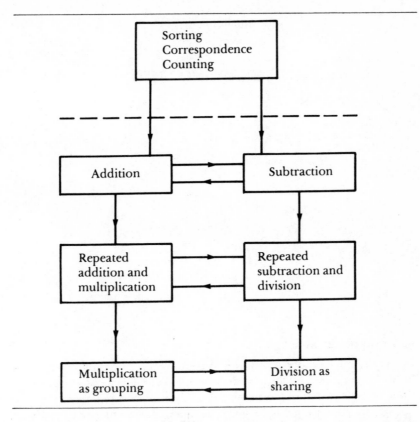

It is not our intention to discuss age norms, because we are concerned with what *all* children should be able to do, given that they do not have difficulties that need special educational treatment.

Children who lack the skills and insight implied by these first two chapters are not likely to get very far in formal arithmetic. They may, with determination on the part of their teachers and themselves, make some progress towards numeracy, but they will never use the language of number with ease and grace.

Many children will have reached this stage of number knowledge by the age of eight, but some will only just have made it while others will have been ready to go ahead for some time. Some may yet be unsure. This, we maintain, is an inescapable fact, and it is up to primary education to cope with it.

CHAPTER 3

THE DENARY NOTATION

Base and place value

It is common practice when counting to make structured collections. Bottles of wine are usually packed in cases of 12 and so we buy large quantities by the case. If we have counted 15 cases, we do not then count the bottles; we use arithmetic to give the result. A wine merchant probably learns such products as facts by sheer repetition in use, and we must always be on guard against allowing these local skills to be taken as norms.

In keeping a tally, one usually marks off in groups of five, using four strokes with the fifth slashed across them. The total value of

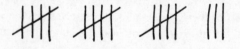

is much easier to recognize as eighteen than

since most of us can see 'three lots of five, add three' more quickly and accurately than we can count the vertical strokes.

Our number system collects in tens. The system is seen in the simplest form in the Egyptian hieratic, the priestly script developed from hieroglyphics. This used / for units, ∩ for tens and ℓ for tens of tens. Thus, 'two hundred and thirty four' could have been written

or

The order does not matter. What does matter is that after nine units are counted and recorded by the stroke symbol, the symbol is changed for one representing a group of ten. We then count in tens and units, hundreds, tens, and units, and so on. We have already noted that the number words themselves, by and large, reflect this pattern. The number ten is then called the *base* of our numbers, which are said to form a denary system.

The history and palaeography of number systems is a topic that many children find interesting. It goes with the history of writing and language generally as one of the areas on which an educated man may be well informed. But it is not to be seen as essential to number skill. One should realize that counting in tens is probably a convention only because we have ten fingers; and numeracy does not demand more knowledge than this of the story of the numbers.

It does, however, demand a clear understanding of the place-value notation that makes modern arithmetic possible. The appearance of this notation in Europe was one of the central technical innovations of the Renaissance, but the complicated history of its passage westwards from India and its perfection by the Arabs *en route* is not relevant and we shall take the notation as given. If teachers find that a historical approach is of help in describing the notation, they will of course use it, but we are concerned here with the final form as it is used, and the next paragraph outlines what must be known and understood. Without this understanding, computation becomes a chain of unexplained and irreducible procedures.

There are only three distinctive features of the system. Most people know them by implication, but we would aim at explicit statement. Compared with the Egyptian denary system ours has these features:

1   In place of the tally count using a unit sign, each of the number words from one to nine is represented by a distinct symbol.

2   In place of tally symbols for groups of ten, the same number symbols are used, but written in a position to the left of the units symbol. Groups of hundreds are shown by writing to the left of the tens. The process is repeated for thousands, tens of thousands, and so on indefinitely.

3   If a count does not include units or one or more of the other groups, the vacant position on the record is marked by entering a special symbol, the zero, as an indicator.

No other symbols are used. It follows that the order of the symbols is significant, and that, since the size of the counting group is given by the position of the symbol, the omission of a zero misplaces each of the symbols to the left of it.

The direct physical model for this is the abacus, a very ancient device whose use probably suggested the notation in the first place. A convenient form is the spike abacus as in the diagram, which is set for 3204. This is the best form for teaching purposes.

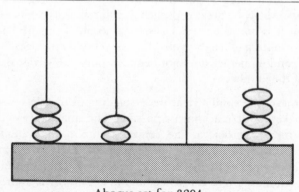

Abacus set for 3204

Beads are put on the wires, and stand for units, tens, hundreds . . . reading from right to left. When the total of beads added to any one spike is ten, they are all removed and replaced with one bead on the next wire.

Although children are no longer taught to compute on them in these countries, forms of the abacus are in daily use in Russia, China, and Japan.

In schools the notation is often written with column headings to avoid confusion

H	T	U
5	6	3
4	0	1
2	7	0

There can be no objection to this, as long as it is done for this reason after the notation is becoming familiar. The first stages should not use numbers larger than the children can arrive at by counting objects.

It is a great help in the classroom, allowing interestingly large numbers to be arrived at without unit counting, to use structured apparatus such as denary number blocks. Here unit cubes are arranged as sticks of 10, sticks of ten as squares of 100 units, and so on.

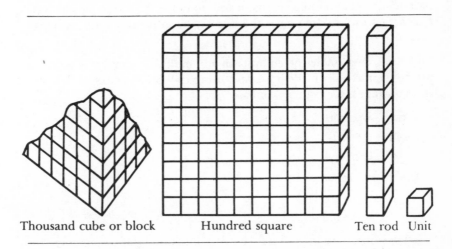

Thousand cube or block   Hundred square   Ten rod   Unit

Sets made in wood or plastic can be bought commercially, and we are convinced that use of such material in childhood is of lasting value in keeping the abstract number system firmly based on activity and experience.

Children used to such apparatus have an immediate mental reference when given a four-digit number, a circumstance of great value in beginning arithmetical computation.

Many other forms of apparatus are available and will be familiar to teachers. Much of it can be improvised or even made up by the children. Milk straws can be cut up into shorter lengths and bundled into tens with elastic bands, or beans can be put into plastic bags. We are not concerned here with the details of teaching, only with the important principle that early number work must, for later success with the largest possible number of people, be done with materials and objects to which the language of number is naturally applicable.

A note on multibase arithmetic

Since the use of ten as a base for counting and enumerating is arbitrary, any other number can be agreed on as the exchange value from column to column. With one possible exception, bases other than ten have no general practical uses, but many teachers argue that working in bases other than ten gives children a greater understanding of place value. Others are of the opinion that it is unnecessary, that, if work in the denary system is properly done, any extension to nondenary counts can be understood very quickly if they happen to be needed. Others may even consider that the work is harmful except to the able, in that it adds to rather than clears up any confusion that the children have about the column notation and exchange values.

There has been sufficient enthusiasm for multibase number studies to justify commercially made apparatus using rods and squares of different denominations, for example, in base 5:

1 rod   = 5 units
1 square = 5 rods
          = 25 units
1 cube   = 5 squares
          = 25 rods
          = 125 units

In schools that have it, one can see the children using this apparatus with ease and confidence, manipulating the small piles of wood or plastic and expressing the results in numerals.

One can also see how a child who has learnt to call the symbol 23 'twenty-three' needs to be at least not below average ability to call the same 23 in a different context 'two three' and realize that it stands for 'eleven' in base 4. There appears to be no experimental evidence to show whether in the long run multibase arithmetic helps or hinders – and by 'in the long run' we mean in the adult making his way in the world. The teacher will need to make a decision appropriate to the children. It should be part of the professional course given to every student intending to teach in a primary school, but this is a specific need and we certainly do not include skill in handling nondenary bases as appropriate to general numeracy.

One is tempted to make an exception with the binary system, which changes column to a count of two and hence uses only the symbols 0 and 1

one	1
two	10
three	11
four	100
five	101
six	110
seven	111
eight	1000
nine	1001
ten	1010

This is, indeed, appropriate to any mathematics course, but one must of course reject the argument sometimes heard, that it is needed for computer operation and should, given the ubiquity of computers, be part of general education. Perhaps we should, as educated and numerate citizens, know what computers can or cannot be used for by their operators and programmers, but binary arithmetic is irrelevant here; it is merely the link between the logic of mathematics and the logic of the electrical circuits that make computerized calculators possible.

If it has any justification in being included in the socially essential topics, it is because it models in number symbols the yes-no, on-off, either-or binary situations that may concern us if we are filing clerks, electricians, or lecturers in logic. Since we do not often need to use such a model in our daily lives, binary arithmetic seems to be an intellectual luxury rather than a prerequisite for living, where we answer yes-no questions or manipulate on-off switches without any need to be aware of their abstract logical structure.

In what follows, all computation or further discussion of number will be restricted to the denary-based arithmetic of our familiar number system, whatever use may be made of other bases as part of the teaching process.

Summary

This chapter can be summarized by giving a flow diagram linking the topics of its development, taking the horizontal dotted line as before to mark off the work of a previous chapter.

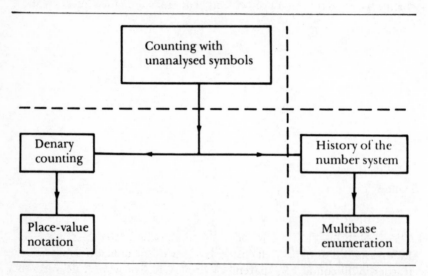

The vertical dotted line is used to separate on the right the two topics which might well appear in mathematics courses, but which are not taken as basic and essential for numeracy.

CHAPTER 4

COMPUTATION WITH WHOLE NUMBERS

The Four Rules: a survey

The third of the Three Rs is commonly understood to consist of the Four Rules, and this term by itself enshrines many of the difficulties that face the teacher in his contact with parents or employers. The word 'rule' suggests a method which has the force of a natural law, and this indeed expresses what many people think about the methods they were taught at school. These methods for obtaining results by calculation are sometimes called 'algorithms'. They have been devised for all the processes of arithmetic, and from now on we shall be discussing them.

Consider the multiplication together of the two numbers 95 and 57. Here are four methods of setting out the calculation:

	1		2		3		4	
	95		95		95		95 ×	57
	× 57		× 57		× 57		47 ×	114
	665		665		4750		23 ×	228
	475		4750		665		11 ×	456
	5415		5415		5415		5 ×	912
							2 ×	1824
							1 ×	3648
								5415

They all give the same answer. It is not likely that the reader uses the last method, and he might have difficulty in explaining why the line 2×1824 was deleted before adding the doubled terms on the right. We only give it to show an algorithm not usually taught. The first three use the same mathematical properties of number and merely rearrange the steps.

If they are shown in turn to an older person who learnt his arithmetic between the First and Second World Wars, one soon sees that they are not taken as equivalent. The zero as the last digit of the product 4750 in (2) would be inserted as a matter of course by a modern teacher who would want his pupils to realize that it records 50 lots of 95. The older citizen, who uses (1), may reject this: 5 times 95 is 475. Asked why he moves the digits one place to the left he is likely to reply: 'Because the 5 is to the left of the 7.' He will usually get the right answer quickly, and is mildly pleased with his skill.

Shown (3) with the suggestion that it could be the best method because the first line written down is the first approximation to the result, he may fail to follow the argument. He might not accept that to arrive at an approximation quickly could be a better test of numeracy than the eventual production of the exact answer.

Teachers of mathematics at all levels are aware of topics which are 'sticking points' in understanding. In elementary arithmetic, given that the number bonds are known, addition is usually learned without serious difficulty. Subtraction, the inverse process, is the first of the 'sticking points', and is still a matter of dispute among teachers who opt for one or another of the 'methods'. Studies have been undertaken purporting to show which method gives the greatest accuracy in numerical arithmetic. Over the last twenty years or so, the 'decomposition method' has tended to take over in primary schools, for reasons which will be discussed later in this chapter. The result is that children transfer to secondary schools whose teachers themselves may do subtraction by 'equal addition', as explained later. Whether it is some children or some teachers who find the resulting situation more difficult is not certain, but secondary schools have been known to ask their feeder primaries to adopt a standard procedure.

A fully numerate adult should be able to understand any proposed method, and to use it confidently enough after a few practice examples. He should no more be thrown by meeting a new method than a fully literate person is at meeting Elizabethan spelling or the Gothic long 's'. The problem arises, no doubt, because children transfer to secondary schools without a secure grasp of whatever method they have been taught: attempts to help them by a different method simply confuse. They are like timid learner-drivers driving on the left in Great Britain, who find themselves in the United States and expected to drive on the right.

The second inverse process, division, appears as less of a problem because skill in it is less frequently demanded. A would-be apprentice who cannot add or subtract is stuck long before he reaches a point at which he needs to admit his ignorance of division. Moreover, in practice, division by a small integer is all that is wanted; the use of large divisors is likely to be over the borderline that separates the general from the special skill. It is notoriously difficult to teach long division with understanding in the very condensed algorithm traditionally used in schools. Those who need and use division frequently, pupils who go on to work with science, engineering, or statistics always used logarithms or slide rules until the appearance of desk calculators began to displace them.

The more recent advent of the cheap integrated circuit calculator introduces a new factor into any discussion of the Four Rules. Their impact on the commercial economics of putting arithmetic to use cannot be denied. There are many who argue that lengthy arithmetical computation is now obsolete, and is done in school or demanded by training officers for the mere sake of doing it. We do not take this extreme view, but accept that the electronic calculator has come to stay. The matter is discussed in Chapter 8. Perhaps the calculator stands to numeracy as the typewriter to calligraphy. We do not now teach copperplate for standard commercial correspondence, but we still expect people to write legibly. The five sections that follow will not mention calculators, but we are aware of their existence in the background. There are five separate sections to this chapter for purposes of exposition, but we should expect the work of these to go on in parallel in the classroom. Arithmetic, and mathematics generally, has suffered much from the convenience of developing one topic at a time. Chapter 2 has already suggested a more integrated teaching approach.

Learning the products

If we assume that products with zero and unity, once understood, are remembered without difficulty, then there remain only 36 products which need to be learnt for the purposes of arithmetic. The table on page 19 is repeated here for convenience, adding the row of zeros which give the products of zero:

×	0	1	2	3	4	5	6	7	8	9
0	0	0	0	0	0	0	0	0	0	0
1		1	2	3	4	5	6	7	8	9
2			4	6	8	10	12	14	16	18
3				9	12	15	18	21	24	27
4					16	20	24	28	32	36
5						25	30	35	40	45
6							36	42	48	54
7								49	56	63
8									64	72
9										81

The 36 required products are shown in the heavy triangle. During the nineteenth century, these products developed a ritual significance paralleled only by the Catechism and the Ten Commandments. As 'The Tables', much of this aura has survived. There are still schools where they are said daily; they are still the subject of parental confrontations; there are still adults for whom ability to recite the 'seven times table' is the test of arithmetical skill. We shall examine these products in detail.

At one time, in the days before decimal currency, reckoning in dozens was common in trade and knowledge of the products of twelve of daily use. Since one passed eleven on the way to twelve, the table of elevens was included, but caused little difficulty because its pattern was obvious at a glance, at least as far as 9×11.

$1 \times 11 = 11$

$2 \times 11 = 22$

$3 \times 11 = 33$

.

.

.

$9 \times 11 = 99$

If these tables are being studied because of their patterns, well and good, but the products are no longer needed for purposes of computation, and need not generally be learnt by heart. Like all number facts, they may be acquired incidentally.

No doubt a few people, or a few hundred thousand people, will continue to work with dozens, but the remaining fifty million will not. Those who need the products of twelve will soon learn them by use. This statement is not dismissive. At one time, one of the contributors to this book did some elementary statistics involving the squares of integers. He now knows by heart, simply by using them for a few weeks, the squares of the numbers from 11 to 20.

We also omit, as part of a deliberate teaching strategy, the table of tens. In the past, learning the products up to 12×12 obscured the quite fundamental issue that up to 9×9 we are dealing with facts that can only be established by counting or addition, but from 10×10 onwards we have results which can be established by computation, by algorithms that depend on the notation. Here multiplication by ten is the basic operation on which everything else in denary arithmetic depends, and we devote a separate section to it.

Chapter 2 has already discussed learning the first 22 products – or 13 if the operation $1 \times$ is not counted. That is, 13 of the 36 basic products should already be known in both random and numerical order as a preliminary informal study not involving computation. There remain 23 products, such as $6 \times 9 = 54$, to be assimilated and ready for instant recall, less numerous, that is, than the grammatical forms of *der* and *ein*, the German equivalents of 'the' and 'a'. It is possible that difficulties are injected into the learning process by the traditional procedures of setting up and teaching 'the tables' by daily repetition in numerical sequence. Some adults who learnt products in this way do not recall them instantly, but recite parts of the table till they reach the required pair.

A more successful strategy continues the product formation of Chapter 2, but now going beyond the maximum of 18. Deriving the products for the first time by using beans or counters now becomes less straightforward because children are dealing with larger groups and can easily miscount or displace the articles. Teachers may well prefer to go over at this point to one of the many forms of structured apparatus available commercially. Such material is usually supplied with handbooks of suggestions for use, and various teacher groups have discussed criteria for assessing their value. But however we deal in practice with the larger numbers now being handled, the general method remains valid. Using 6 and 9 as examples, the child assembles or otherwise represents 6 lots of 9, determines the result, and records it as

$$6 \times 9 = 54.$$

The product is now transposed to give

$9 \times 6 = 54$

and the child tries to learn the result. The kind of test shown on page 33 both assesses progress and provides consolidating practice.

As the facts slowly accumulate, the child can begin to list them systematically in as many ways as possible.

Examples

1

×	1	2	3	4	5	6	7	8	9
7				28			49		

The table for '7 lots of . . .'

The table should be drawn up carefully by the child, using squared paper as a guide. If he cannot fill in all the spaces he should be referred back to the apparatus or the counters, not merely be told what to put.

2

0×7	$= 0$
1×7	$= 7$
2×7	$= 14$
3×7	$= 21$
4×7	$= 28$
5×7	$= 35$
6×7	$= 42$
7×7	$= 49$
8×7	$= 56$
9×7	$= 63$

This is the table of sevens written out as before. Having done it neatly on a slip of wide-lined paper provided, the child can be asked to cut it down the dotted line and then along all horizontal lines to give 20 pieces. These are now to be assembled in the correct order.

The above are given as examples of the approach to product learning that we recommend. Progress and practice exercises as on page 33, not necessarily set explicitly as 'tests', ought to continue, although we feel once a child begins to get them all right they should be set less frequently. By then, the pupil will keep them in mind through using the products in the next stages of his learning programme. A final check-test that can be set at any time is to issue each child with a 10 × 10 blank square. The top left hand square is left blank or entered with a multiplication sign. The first row and the first column are entered with the numbers 1 to 9 dictated in any random order; the class then have to fill the 81 remaining squares with the correct products.

×	9	2	4	5	3	7	1	6	8
3	27	6							
5	45								
8									
1									
2									
7									
9									
6									
4									

Random order product square

As they become known the products can be rewritten as quotients, exactly as on page 32. Each pair of numbers now gives rise to four related facts, of which only one needs to be discovered by counting and grouping.

Using our example $6 \times 9 = 54$

$9 \times 6 = 54$

$54 \div 6 = 9$

$54 \div 9 = 6$

It is useful to give children products and ask them to express them as factors: if this is done properly, it also introduces the term 'factors' in context, that is, the numbers which when multiplied together give the required number.

The work should not be rushed, and it is not necessary for all children to have learnt the products by heart before going on to do long multiplication. A neatly written product square can be used for reference, just as later one uses logarithm tables. Where this is permitted, the product activities and evaluation tests should continue until a random order square can be filled in without hesitation. The ability to complete rapidly both this and a similar square for the addition facts, running along the rows entering the sums or products, is almost a guarantee that the basic number facts needed for computation are properly known.

Older readers who learnt their products by repetition of tables, who have always regarded knowledge of 'The Tables' as essential, will on consideration agree that nobody uses the tables as such, any more than an amorous Roman soldier ever recited *amo, amas, amat* .... What one does use and must know are the products as they arise and are needed, which is at random. We might well arrange the products in tabular form to investigate the many number patterns that their structure produces, but this is for purposes of study and not for learning, although of course it may assist learning considerably. Many less able children can only recover in sequence the products learnt in sequence, going through all or part of a table before recalling the product they need.

The very important commutative property of multiplication was traditionally ignored. Indeed, some older teachers in the past seemed to be unaware of it. They would, for example, range the tables in order of difficulty, with the 'two times table' first and the 'seven times table' last. Whichever table is left to last should be the easiest of all, since it only contains one new fact. In the table of sevens it is $7 \times 7 = 49$. All the other products of seven are recorded in the tables of twos, threes . . ., but the mental process of sequential learning obscures this. It is interesting to note that music makes a quite different kind of demand. One recognizes a tune as a whole, not noticing a short sequence of notes in it that may occur also in a different tune. Table learning burdens some people for life with the need to isolate facts from an artificial assembly of them. The more able do this without realizing the skill they are bringing into play, the less able do not and fail to achieve instant recall.

What we are recommending in place of table learning is more demanding on the teacher, and we hope more rewarding to the pupil. But there can be no relaxing on the content of the final skill. Children *must* learn the number facts.

Addition

We have at last reached the point at which we can begin to discuss computation as a process that draws together the extensive preliminary skills and knowledge that arithmetical teaching so often takes too sketchily and too fast. If children have already become familiar with the denary notation through the use of structured or extemporized apparatus in counting and making tallies, the algorithm for addition should present little difficulty. The bead abacus shown on page 38 is probably the best single piece of equipment. Its use often comes as a revelation to adult student teachers whose own understanding of notation is not completely confident.

Adding one bead at a time to the units wire of an abacus is equivalent to addition by counting. The child already knows that when the count reaches a total of 10 the unit beads are removed and replaced with a bead on the tens wire. Addition now becomes a matter of replacing a digital count by the number bonds, known to a limit of 9 + 9. The traditional algorithm that gives the sum 28 + 36 = 64 is set out in two forms which differ only in placing the carrying figure.

```
  28        2 8
 +36      +3₁6
 ───       ───
  64        6 4
  ₁
```

The mental process, at first vocalized by the teacher and pupils, is 'eight and six, fourteen. Put down the four and carry one. Two and three, five; and one six.' Adults who add up at all frequently are scarcely conscious of this mental process, but will produce it, or something very like it, if asked to explain carefully how they get the result.

Since we are aiming at maximum understanding in the early stages, when the less able child may be acquiring a technique of short cuts that eventually lets him down, we recommend a more precise description of what is being done.

One does not, of course, 'put down four and carry one'. It is *ten* that is carried. It helps many children to do the addition in two stages

```
  28
 +36
 ───
  14

  50
 ───
  64
```

Here we say 'eight and six, fourteen' and the 14 is set down. 'Twenty and thirty, fifty', and this too is written. Then the two parts are added without difficulty. Eventually this 'extended algorithm' is condensed into the usual form, and the child can say:

> 'Eight and six, fourteen. Put down the four and carry one *ten*. Twenty and thirty, fifty, and one ten sixty.'

This takes a little longer, but it does not risk becoming a meaningless rigmarole. It may be contracted into the first form of words by an able child on his own initiative, but the aim eventually is that it should be submerged altogether. The numerate person does not add up to a background of vocalization, any more than the literate person vocalizes his reading of a newspaper.

Nevertheless, the child learns to read by vocalization and he learns to compute in the same way. The vocalization is an essential stage, and it must, we believe, be done as exactly as possible, to help as many people as possible in the long run. It is this point that makes it important. An adult who reads without difficulty has long forgotten how he learned, yet which method is most likely to help the slow reader is a matter of educational importance. We believe that a careful verbal statement of the stages in arriving at a sum, with the use of simple apparatus to model place value in the early stages, leads to operational efficiency with understanding. The operation is not efficient if the column totals are reached by counting on, touching the fingers, or putting dots on the paper, and two-column addition should not begin before the bonds are thoroughly known and checked. The algorithm, once understood, is easily extended to three or more digits.

The accurate addition of a long column of digits to reach a required total is a special skill that develops very quickly in anyone who practises it. Employers who expect school-leavers to do long tots and cross tots at sight are asking too much of general numeracy, particularly where commercial practice now uses machines. We would agree that a numerate person should be able to deal with sums up to $9 + 9$ by instant recall, and could be expected to add a column of figures accurately, but only in his own time. If he can do this, speed comes with practice but only with practice. The small shopkeeper is not being reasonable if he blames the schools when his new assistant cannot add up purchases as quickly as he can. One must distinguish between basic ability and the outcome of the repetitive practice which the worker may obtain in the context of his job.

Subtraction

The methods of subtraction, like the multiplication tables, form part of the sociology of mathematics. Children who do not follow the notational manipulation that makes subtraction possible take refuge in drill routines often irrelevant to a particular example. Since subtraction in one or other of its forms is a common arithmetical operation, it is here that numeracy begins to tell. Because the forms in which it arises are often not those of the set subtraction exercise of the schools, the classroom methods may be inappropriate. They are, nevertheless, the only form in which most people are able to discuss the problem.

Reference to the section *Aspects of subtraction* on page 24 does not yet show what the problem is. We considered $9 - 2 = 7$ as the inverse of $7 + 2 = 9$, where the existence of the sum guaranteed the existence of the difference, and showed how the operation could be interpreted with sets of objects such as counters and sweets. Given the sum $26 + 49 = 75$ it follows at once that $75 - 49 = 26$. The interpretation of this in groups of objects is equally clear, but how do we find the difference without a full count? In addition with more than one digit, the columns are added separately, but in

$$\begin{array}{r} 75 \\ -49 \\ \hline \end{array}$$

the one-to-one correspondence breaks down because 9 is greater than 5. One cannot subtract 9 from 5, even though the subtractor 49 as a whole is smaller than the 75 it is taken from. To deal with this situation within the notation is the problem of subtraction.

Several algorithms are possible, which may become inflexible rules in the minds of those who are taught them. A numerate person adopts the best method in the circumstances, which may be no algorithm as such. Asked to take 2 from 1000, most of us would say 998 without conscious calculation, but some schoolchildren, given

$$\begin{array}{r} 1000 \\ - 2 \\ \hline \end{array}$$

have been known to produce

$$\begin{array}{r} \cancel{1}^0\ \cancel{0}^9\ \cancel{0}^9\ 0^{10} \\ -\ 2 \\ \hline 9\ \ 9\ \ 8 \end{array}$$

In this example the drilled algorithm (the decomposition method as discussed below) has taken over from common sense. Thus we teach innumeracy.

Argument has gone on for years about the relative merits of the two methods commonly taught, decomposition and equal addition. We shall discuss them briefly, but only for the insight they give into the problem of handling the notation in arithmetic and of dealing with the situations in which number skills are called for.

The decomposition method is the one most easily demonstrated with the abacus or the common structured apparatus, and for this reason tends to be most commonly found in primary schools today.

Consider   75
            −49

If the 75 were set up on an abacus, there would be 7 beads on the tens wire and 5 on the units wire. Since one cannot remove 9 beads from the wire, one naturally takes a bead from the tens wire, leaving 6, and changes this back for 10 unit beads, that is, we *decompose* the ten into its units. We now have 15 unit beads, and the direct subtraction is possible. Representing the bead operations in the notation gives

$$7^6 \ 5^{10}$$
$$\underline{-4 \ 9}$$
$$2 \ 6$$

One finds trivial differences in setting out the calculation, but this is the principle at work.

Once children have got used to it they subtract accurately enough, but one can see two disadvantages:

1   It tends to be used regardless of the numbers concerned. The example $1000 - 2$ is extreme, but the method is similarly inappropriate for $27 - 9$.
2   It not only requires paper and pencil, but tags the original digits with subsidiary figures and deletions. In some applications, such as list price less discount, this means that calculations need to be written out separately.

The equal addition method is the mathematically respectable form of 'borrow one, pay back the one'. It is the method used by a large majority of older people, for many of whom it is merely a routine that produces the right answer. It consists in recognizing the equivalence of the two expressions $75 - 49$ and $(75 + 10) - (49 + 10)$. Ten is added to both, and hence its name. The difference, of course, is unchanged. In the notation, the 10 units are added to the units column of the 75 to make it $(75 + 15)$, and one ten is added to the tens column of the 49 to make it $(50 + 9)$.

Then $75 - 49 = (75 + 10) - (49 + 10)$
$$= (70 + 15) - (50 + 9)$$
$$= 20 + 6$$

This can be set out as

$$
\begin{array}{r}
7\,5^{10} \\
-4\,9 \\
\hline
1
\end{array}
$$

Here the ten added to the minuend (the number from which we subtract) is written as such, then the other ten as 1 in the tens column. Writing it as

$$
\begin{array}{r}
7\,^{1}5 \\
-4_{1}\,9 \\
\end{array}
$$

has produced the infamous rigmarole 'nine from 5 won't go, borrow 1, pay back the 1'. Note, too, that the first form does not actually need the number fact $15 - 9 = 6$ since the result is suggested in two stages: 9 from 10, 1; and 5 is 6.

The main objection to this algorithm is the failure to teach it except as a routine. It is less easily demonstrated with apparatus, but perhaps makes better use of the notation once it is known. It can scarcely be taught with understanding except by using apparatus, since it is clear that the demonstration given is not accessible to small children. A suitable verbalization might be '9 from 5 I cannot, add 10 top and bottom; 9 from 15, 6; 50 from 70, 20'. As before, it is a written process difficult to perform mentally.

As soon as one leaves the classroom, however, one notices that

1   most 'subtractions' are in fact the giving of change after purchase.

2   subtractions, when needed, usually call for an immediate mental calculation.

These points are important, because the standard school procedures ignore them. The traditional arithmetic book insists on regarding expenditure as subtractive, isolating the process from what we actually *do*:

'Paul has 50p and buys some sweets for 16p.
How much has he left?'

It is obvious that if we try this, by giving a child 50p and taking him to a
sweet shop, he will not call for pencil and paper and produce the answer by
algorithm. Nor will the shopkeeper. Paul may give him the exact money,
but if he does not the shopkeeper will put down the sweets to the value of
16p and build up this amount to the value of the coins tendered. The
shopkeeper says '16 and 4 is 20, 30, 40, 50', putting down the coins as he
does so. The correct change is 34p and this is now on the counter, but
nobody has calculated it. If Paul has followed the process, he will pick up
his change in confidence without knowing that he has 34p.

This process, which in one form or another is universal where change is not
given by an automatic till, is avoided completely in schools; probably
because it does not reduce to an algorithm and is done according to the
coins available. The shopkeeper could well say 'I haven't any coppers.
Have you got a penny?' Paul will then put down 1p if he has one and the
transaction will go on as if the sweet cost 15p. It is this easy familiarity with
a particular set of numbers, the values of the coins of the realm, whose lack
the retail trade rightly deplores in its young entrants. Possibly the answer
lies with the shops as well as in schools. Could not local chambers of trade
cooperate with colleges of further education to run one-week courses at the
end of the school year, planned for those leavers who want to work in
shops?

We often adapt this process *ad hoc* to deal with such subtractions as we need
in daily life. A teacher has a class of 36 and 19 of them are withdrawn on
that day. How many school meals are needed? She is likely to think sotto
voice '19 and 1 is 20, and 16 is 17', making the step from 20 to 36 without
analysis. As before, we cannot reduce this mental activity to an algorithm,
but this is what a numerate person should be able to do.

There is, indeed, a third method of subtraction sometimes taught which
approaches these practical methods much more closely. It consists of mentally transforming the subtraction bond to a complementary addition. The
exact mental process is best shown with the box notation. Instead of trying
to complete the statement

$$9 - 4 = \Box$$

by filling the correct number in the box, one completes

$$4 + \Box = 9.$$

Set out in the usual way the two are identical

$$\begin{array}{r} 9 \\ -\,4 \\ \hline 5 \end{array}$$

A behaviourist psychologist might see two people each of whom wrote the 5 without hesitation, but the second, using complementary addition, would say to himself '4 and 5 is 9', writing the 5 as he says it. If children are taught their number bonding as recommended, with extensive use of all the available methods of recording, check-testing and practising, it needs only a bias towards statements such as $7 + \square = 15$ to give them the needed skill. They then deal with, e.g., $79 - 36 = 43$ in the usual way, but using the mental phrase 'up to' for as long as they need its help. $75 - 49$ is now dealt with mentally without using subsidiary numerals.

$$\begin{array}{r} 75 \\ -\,49 \\ \hline \end{array}$$

One thinks of 75 as sixty-fifteen, and then it is '9 up to 15, **6**; 40 up to 60, **20**', writing the difference 26 as the digits (in bold) are obtained. It is true that it takes longer for the children to become proficient, but once they are proficient the process is quicker and neater. We would not actually teach children to drop the words 'forty', 'sixty' and 'twenty' in the example, but as they become adept they will no doubt do so. Teachers themselves working with the older children on blackboard can begin to do so. The final contracted vocalization is, for $75 - 49$

'9 up to 15, 6; 4 up to 6, 2'

It is now understood that the second phrase deals with the tens column. A person who does an occasional subtraction will continue to use the mental prop, but anyone who uses differences daily will soon find the process automatic. At first, children may find it easier to use a 'half-way house'. In the example $75 - 49$, one could choose 50 and say '49 up to 50, 1; 50 up to 75, 25; that's 26 altogether'.

Since the process parallels what one actually does with coins in building up change on a counter, and since it relieves the algorithm of its outcrop of diacritical digits, we would like to see it taught more often. This method of dealing with subtraction by 'complementary addition' is sometimes called the Italian method, from its origins in Florentine commerce. It would certainly be worth while for secondary schools to do a short intensive transition course during the final year, and the outcome would be a much larger proportion of school-leavers that come much closer to full numeracy.

Multiplication and division by ten

In our denary notation, multiplication by 10, written using the symbol for one and the place holder zero, gives a result which has been one more tripping point for the child struggling along the path to numeracy. It is almost certain that the schools are to blame; if not the schools of the children then the schools of the parents. Two lots of ten are twenty, and in the notation we write this 20. Treating this for the moment as an 'atomic' number fact in our table of products – although on page 47 we argue for its omission – we get

$$2 \times 10 = 20$$
$$10 \times 2 = 20$$

The pitfall is now ready. Ten lots of two is 20. Ten lots of three is $10 \times 3 = 30$.

The completely misleading and erroneous rule now invites formulation. We give it only to warn against it.

First spurious rule: to multiply by ten, add a zero.

This is the disaster point. The child is now well away along the wrong road, furnished with a clear and distinct misconception, and likely in the end to join the ranks of those who 'can't do decimals'.

The difficulty must be faced squarely: it is one of those not uncommon situations in mathematics where the simple intuitive approach proves inadequate. The statement $7 \times 9 = 63$ is much more difficult to learn than $2 \times 10 = 20$, but whereas the first is an 'atomic' fact that cannot be analysed by a standard computational algorithm, 2×10 is the first computed result that arises in denary notation. It is constructed from the two atomic facts $2 \times 0 = 0$ and $2 \times 1 = 2$, combined together in the place-value notation. Ten itself is a compound symbol, and must, for the purposes of computation, be understood as such, even if we recommend as on page 18 that in the first stages of number learning it should not be analysed.

Two lots of ten, represented by groups of counters, gives the fact $2 \times 10 = 20$ by exactly the same process as counters can yield $7 \times 9 = 63$. If we use an abacus which represents the notation the difference becomes clear.

For 10, we have one bead on the ten wire and none on the units wire. To form 2×10, we simply double up on the tens wire. For 2, the ten wire is empty, and for 10×2, each bead on the unit wire becomes a ten bead and

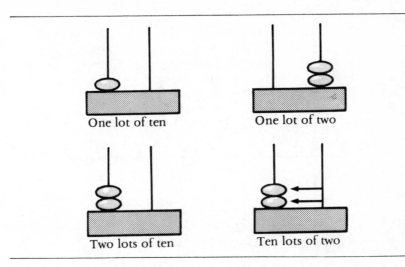

One lot of ten

One lot of two

Two lots of ten

Ten lots of two

shifts across to the ten wire. We use only two beads to determine the product, not twenty. It has been *calculated.* The actual operative form of the rule can now be seen.

The correct rule:

> To multiply by ten, shift the digits one place to the left, entering a zero in the empty column left on the right.

If the multiplication is set out as it should be, with all figures in the correct columns, the distinction is clear but less obvious

T	U		T	U
1	0			2
×	2		× 1	0
2	0		2	0

The correct rule should not be taught by rote, but formulated as the children consider examples. The examples cause no difficulty if, but only if, they are done using an abacus or its equivalent. As with many mathematical processes, the simplest examples such as 10×2 show the rules less clearly than a more complex one, as in the abacus setting that follows for $10 \times 236 = 2360$.

There are few better illustrations of the value of structured apparatus in teaching arithmetic. The abacus wires are there whether they have beads on or not, and to set up a number on them, to demonstrate an algorithm with

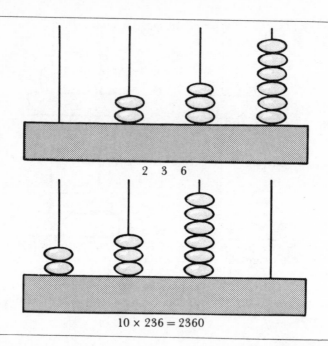

2 3 6

$10 \times 236 = 2360$

their use, makes it necessary to decide on which wires the beads are to be put. Only a child whose introduction to the notation is clearly anchored in the place value of digits in vertical columns can avoid the misconceptions that arise in arithmetic.

Many teachers, as we have said, consider that a study of multibase arithmetic, using bases other than 10, is the best way to avoid later confusion. Our own brief, however, is to consider the final content of arithmetical knowledge that is needed for adult social and economic adequacy, and we omit multibase studies as a teaching method whose relevance to final numeracy is not certain.

It should follow at once that the division facts corresponding to

$$2 \times 10 = 20$$
$$10 \times 2 = 20$$

are

$$20 \div 10 = 2$$
$$20 \div 2 = 10$$

Once more, the abacus makes quite explicit what is happening:

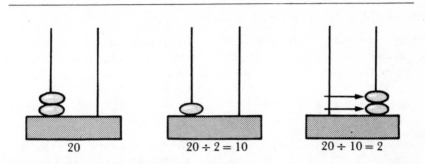

| 20 | $20 \div 2 = 10$ | $20 \div 10 = 2$ |

Division of numbers by 10 shifts them one place to the right, the inverse of the process that originally shifted them one place to the left. It follows that a number is only divisible by 10 if it ends in a zero, an empty column into which the penultimate digit can shift. This section does not consider numbers such as 12 which are not divisible by 10 in the sense of Chapter 2.

Multiplication

Pupils are now, but only now, ready to tackle multiplication, whose standard algorithm, taught too early, has caused so many later problems. When children demonstrably get their arithmetic right, it is not easy to blame their teaching for hang-ups that occur subsequently. It is the failure to discuss multiplication in terms of place value that results in problems with decimal fractions when this essential topic is being taught. Multiplication, if it is to be used at all, needs to become automatic, but to become automatic in use is very different from being a meaningless routine at the outset.

We are not now examining techniques for teaching multiplication. It will be enough to analyse the algorithm in terms of the denary notation. The example of an algorithm we shall take is

$$
\begin{array}{r}
257 \\
\times\, 23 \\
\hline
771 \\
5140 \\
\hline
5911 \\
\end{array}
$$

This works because of several arithmetical and notation facts.

1   23   = 20 + 3
2   257 × (20 + 3) = (257 × 20) + (257 × 3)
3   20   = 2 × 10

Of these three statements, the first is notational, the second is a mathematical law, here made explicit but tacitly assumed in practice (the so-called distributive law which tells us how to use brackets), and the third is the basic computation discussed in the last section. The second tells us in effect that we can multiply by 3 and by 20 separately, adding the two products to get the final product 5911.

Multiplication of 257 by 3 is itself a combination of number facts, here $3 \times 7 = 21$, $3 \times 5 = 15$, $3 \times 2 = 6$, together with operations arising from the denary notation. In this example

```
        257
        × 3
         21 which is 3 ×    7
        150 which is 3 ×   50
        600 which is 3 × 200
Total   771 which is 3 × 257
```

Multiplication by 20 similarly uses the facts $2 \times 7 = 14$, $2 \times 5 = 10$, $2 \times 2 = 4$, together with the operation 10 × that displaces all digits to the left, so that we have

```
         257
        × 20
         140 which is 20 ×    7
        1000 which is 20 ×   50
        4000 which is 20 × 200
Total   5140 which is 20 × 257
```

Addition of 5140 and 771 gives the total product 5911. This dissected algorithm is usually contracted into

```
   257
    23
   771
  5140
  5911
```

Older people may omit the zero in the second line (the second partial product so-called), but most teachers would not now permit this.

A numerate person should be able to follow such a dissection of any algorithm known to him. The extent to which it should be discussed in schools, or actually used as a teaching technique in dealing with multiplication, is a matter of opinion. Certainly something more than a step-by-step set of instructions is called for. It will be seen that the key to the correct result is to keep all digits in vertical alignment so that place values are accurately maintained. It is perhaps worth comment that in schools that use lined paper the lines are used horizontally to keep neat rows of figures. If the lines were used *vertically*, the notation would almost look after itself, whatever the horizontal deviations.

```
Th  H  T  U
    2  5  7
 ×  2  3
      7  1
   7  4  0
 5
   1
 5  9  1  1
```

Shopkeepers and accountants know that a common source of error is entering multidigit figures out of vertical alignment.

In the first stages, multiplication of several digits by a single integer should be done using apparatus. This work should continue until children can multiply by one digit confidently and accurately before proceeding. Some sort of extended or dissected algorithm should then be taught for multiplication by two or more digits. The contraction into the usual algorithm should be allowed as the pupils see what is happening. We prefer the algorithm

```
  257
× 23
5140
 771
5911
```

because the first partial product, the first step in finding the first product, is 5140, and this is a first approximation to the final result. Anyone who appreciates the relative importance of the partial products when multiplying any quantity by a number of several digits is likely to be fully numerate.

Summary (i)

Before continuing we summarize this section by a flow diagram.

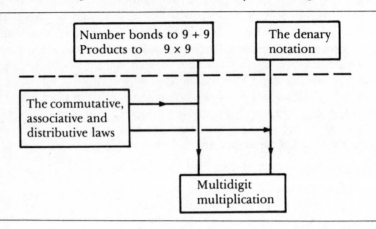

A final word is needed about the formal laws of arithmetic, since any mention of them in a teaching context apparently suggests that they should be taught. This is nonsense. We teach a baby to say Daddy and Teddy without teaching the difference between voiced and unvoiced dentals. Children should become familiar with these rules implicitly by use, not explicitly by exposition.

For reference only we give the laws in their usual form, where $a, b, c \ldots$ stand for numbers:

1 The commutative law   (i) $a + b = b + a$

(ii) $a \times b = b \times a$

2 The associative law   (i) $(a + b) + c = (a + b) + c$

(ii) $(a \times b) \times c = a \times (b \times c)$

3 The distributive law   (i) $a \times (b + c) = (a \times b) + (a \times c)$

Part of the formal importance of these laws lies in the fact that the inverse processes of subtraction and division do *not* follow them.

Division

The simple physical process of sharing or continued subtraction can in principle be extended to groups of any size. Indeed, it is worth noting that in practice this is the commonest way of dealing with a simple situation. How otherwise does one distribute a bag of sweets among children at a party? Not even a mathematics teacher would count the sweets, count the children, and calculate the share of each. The earlier chapter also included the division facts arising as the inverse of multiplication, and on page 49 we suggested the extension of these inverses to products up to 9 × 9, not as separate items to be remembered but indirectly through the products.

Thus we can deal with 59 ÷ 6. We know that 6 × 9 = 54, hence the quotient is 9 with a remainder of 5. We shall write it as

$$\frac{59}{6} = 9 \text{ rem. } 5$$

The mental step is equivalent to not only knowing the statement 6 × 9 = 54, but realizing that this is as close as we can get to 59 without overshooting it. Subtraction then gives the remainder.

A similar process can be used via the notation for dealing with, e.g., 59 ÷ 3. Here there is the ever present danger of replacing knowledge with routine manipulation, although in the end the routine appears to be necessary.

'3 into 5 goes 1 with 2 over, 3 into 29 goes 9 with 2 over'

$$\frac{59}{3} = 19 \text{ rem. } 2$$

This is the correct result and this is the form of words that most people use unless they divide so frequently that they learn to dispense with it. One soon sees that an attempt to be pedantic lands one in trouble. Writing as (50 + 9) ÷ 3 only makes it more difficult to follow. The locution '3 into 50 goes 10 and 20 over' or '3 into 5 tens . . .' seems even worse.

The matter can easily be settled by taking the children first through such calculations either with an abacus or with bundles of straws or bags of

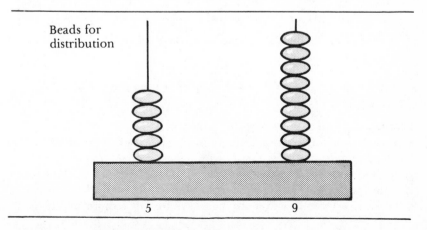

Beads for
distribution

5   9

beans. We shall use an abacus diagram, showing beads worth 59 units available for distribution.

We are to give equal shares of the beads to three people, remembering that each bead on the left is worth ten times as much as a bead on the right. Each person will set up his share on his own abacus.

First
distribution

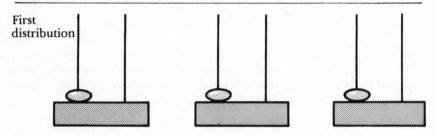

After the first distribution of one ten bead each, there are now only two beads left on the tens wire. These cannot be shared equally among three, so we change them for ten units each. We now have 29 units, and we know as a product fact that $3 \times 9 = 27$. Hence we give the three recipients 9 each, leaving 2 undistributed. Small children like to call them 'teacher's share'. The final picture showing three shares of 19 is

Final
distribution

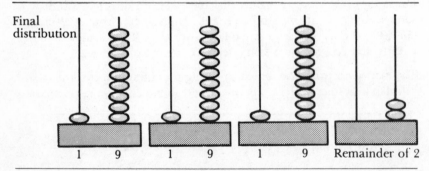

    1   9   1   9   1   9   Remainder of 2

The whole process is much easier to demonstrate on an abacus than to describe.

It is now clear what one is doing with the algorithm. The original concrete experience is modelled with the digits in position, and all we ask is that the children should, always at first but more and more infrequently as they gain speed and accuracy, accompany early division exercises with the use of structural apparatus. The method extends itself without further conceptual difficulty to dividends with more digits. We would recommend the use of the abacus for a few examples using three digits, extending to more than three by computation alone:

$$\frac{67328}{7} = 9618 \text{ rem. } 2$$

Since the digit 6 is not divisible by 7 in this example we go straight on to 67. It would be intolerable to have to think of this as 67,000 every time, and '7 into 67 goes 9' is the best guide to a result. With the apparatus in the background, there is a basis of understanding, however rapidly and mechanically one learns to divide by single digits. Since few of us ever have to do these calculations regularly, we would not regard speed in such processes as a measure of numeracy. Timed tests do not necessarily select the numerate, but may only sort out those who have recently practised the operation tested.

The algorithm may be set out in several ways, with or without writing in the figures carried from column to column. Any one of them should be clear and intelligible if the process is understood. The method depends directly on a knowledge of the division facts as the inverses of the products up to 9×9, and it cannot help if we want a quotient with two- or three-digit divisors. The algorithm for giving such a result with multidigit divisors is the most difficult of the processes to learn, even by rote. It is certainly the most difficult to describe. The fundamental difference is that in this new algorithm we use continued subtraction to arrive at the result.

Consider as an example the physical situation, not of giving equal shares of 93,721 objects to 23 people, but of determining how many people can be given 23 each. Such a situation cannot arise very often, but it is easy enough to imagine. It could be solved without computation by subtracting 23 articles continually until none or less than 23 were left, keeping a tally of the subtractions. If it is seen that the 'long division' algorithm is a notational adaptation of this process, it should not be difficult to understand.

A simpler example is easier to discuss. We have 533 tokens and we want to distribute them in pairs. At once we see that 200 people can be given 2 each, thus disposing of 400. The result for this partial distribution can be set out

```
  200 people receive 2 each
2)533 tokens in all
  400 tokens in first distribution
  133 left for distribution
```

We now see, from our knowledge of the product $2 \times 6 = 12$ and of the positional notation, that we can serve another 60 people, using up 100 tokens. We shall add the record of this second stage to the first, getting

```
      60 people
     200 people
   2)533 tokens
     400 first distribution
     133 tokens left
     120 second distribution
      13 tokens left
```

Finally we can give 6 people 2 each, leaving one. There is no need to repeat all the annotations

```
       6
      60
     200 persons. Total 266
   2)523
     400
     133
     120
      13
      12
       1 remaining token
```

The three separate groups of people, in hundreds, tens, and singles can now be written as a three-digit number, and the whole thing freed from its anchorage in people and tokens. All we need do is leave blank the positions occupied by the zeros in 200, the first partial quotient, so that the digits 6 (tens) and 6 (units) can be put in as they are found. We then get, directly

```
     266
   2)533
     400
     133
     120
      13
      12
       1
```

This is a contracted version which is often contracted still further by omitting the zeros and not completing the subtractions along the entire row of digits. We then get

```
    266
2)533
    4
   ──
   13
   12
   ──
    13
    12
    ──
     1
```

The omitted digits should be checked by referring to the previous layout. This is the process known as long division as usually set out, although some older readers may be more familiar with

```
2)533( 266 rem. 1
  4
 ──
 13
 12
 ──
  13
  12
  ──
   1
```

The example has used a single-digit divisor to simplify the exposition. We can now set out, without annotation, the earlier example of 93,721 divided by 23, in all three forms:

4	4074	4074
70	23)93721	23)93721
4000	92000	92
23)93721	1721	172
92000	1610	161
1721	111	111
1610	92	92
111	19	19
92		
19		
fully analysed	partly contracted	fully contracted

The contracted form is reasonable enough. If (which is hardly likely) large numbers of workers in offices or factories were required to do frequent divisions in writing like this, it would save both time and ink. Note that, as with long multiplication, the success of the algorithm depends on keeping the column alignment that determines the value of the digits: this is less easy with the fully contracted version. One would recommend that the full form be taught and used, to meet the situations (and surely the very infrequent situations) where long division is needed. It is a fairly straightforward matter to go through such examples using base-ten arithmetic blocks, at least with children who are not educationally disadvantaged. Teachers who use the apparatus will be familiar with the techniques.

Although we feel that the algorithm is of marginal importance, particularly, as we shall argue later, against a commercial background of cheap calculators, its understanding is well within what we think of as full numeracy, and we should expect adults to complete a long division, at least in their own time, and perhaps after a 'refresher demonstration' of the method. That so many adults and school-leavers cannot cope, and fail even more when the process is extended to decimal fractions, seems due to a traditional method of teaching that we now discuss.

Many adults have never seen the uncontracted algorithm. They began as children with the contracted version, presented with a verbal commentary that enabled them to get the result. This vocalization is quite familiar to older people, and regrettably, to many of the younger. Using our example of division by 2 to make the point more simply we should have

$$
\begin{array}{r}
266 \\
2{\overline{\smash{\big)}\,533}} \\
\underline{4} \\
13 \\
\underline{12} \\
13 \\
\underline{12} \\
1
\end{array}
$$

The pupil following his teaching, would say '2 into 5, 2; 2 × 2 is 4. 4 from 5 is 1, bring down the 3. 2 into 13, 6; 2 × 6 is 12. 10 from 13, 1, bring down the 3 ...'

This process arrives at the result, but in terms of anything that could be called numeracy it is deplorable unless the person doing it has a clear idea of what is happening. The contracted algorithm 'brings down' the 1 and the 3 in turn, but reference to the full version shows that they are there as the result of subtracting the zero digits in 400. It is even worse in the two-digit example, where we have: 'bring down the 7; 23 into 17 won't go. Put down zero (or nought) and bring down the 2 . . .' Reference to the analysed form shows at once that the first or partial quotient 4000 is followed by 70, not 700.

If this process is not worth understanding, it is not worth doing. If it is not understood, one can almost guarantee a more serious difficulty later with decimal fractions, whether calculating devices are used or not. It is also odd how, in teaching long division, the rule arose that one had to deal with each stage of the calculation by using the largest possible product. As a result, pupils' division exercises were often accompanied by trial multiplications in the margin, often crossed out or abandoned halfway through. Alongside the algorithm for $93721 \div 23$ given, one might see such scribbled calculations as

$$
\begin{array}{ccc}
23 & 23\!\!\!\!/ & 23 \\
\times\, 4 & \times\!\!\!\!/\, 6 & \times\, 7 \\
\hline
92 & /138 & 161 \\
\end{array}
$$

The product 6×23 would have been crossed out because the pupil realized that it was too small. He then does the trial multiplication 7×23 and uses it. This is quite unnecessary. In the analysed form, having got the product $60 \times 23 = 1380$, one can use it without trouble

$$
\begin{array}{r}
4 \\
10 \\
60 \\
4000 \\
23\overline{)93721} \\
92000 \\
\hline
1721 \\
1380 \\
\hline
341 \\
230 \\
\hline
111 \\
92 \\
\hline
19 \\
\end{array}
$$

One merely uses one more stage. Since the pupil's marginalia may well have included the trial subtraction,

172
138
‾‾‾
 34
‾‾‾

which he did to satisfy himself that 138 was in fact too small, the extra step had been done anyway, out of context.

The examples $513 \div 2$, $93721 \div 23$ were chosen to illustrate the process. We repeat that it would not be used in practice for the first quotient, and we recommend that if the method is taught at all, as an exercise in the use of number bonds and the positional notation, four-digit numbers divided by two-digit numbers should be the limit of examples set. If division needs to be done frequently in a commercial or technical context, long division is inefficient and seriously uneconomic; and no employer aiming at efficiency would want it to be done.

Summary (ii)

The topics in this section can be summarized in a flow diagram as before.

The last box of the diagram contains a query because it is not certain that the process is essential to general numeracy.

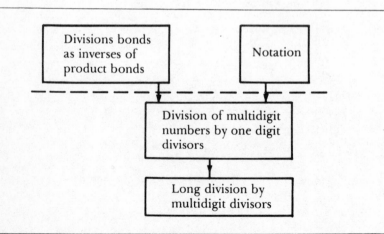

CHAPTER 5

NUMBER EXTENDED : THE FRACTION CONCEPT

Fractions and decimals

In the past the introduction of decimals was invariably postponed for two or three years after children had become familiar with fractions. Many parents will remember having 'done' the Four Rules with fractions before meeting decimal calculations of any kind. Since by then some children were beginning to have difficulties, and since the thorough notation studies advocated in Chapter 3 were rarely attempted, decimals often proved an arithmetical *pons asinorum*. Much of the dismay among older people when the decimal coinage was being debated arose simply from the overtones of the word 'decimal'. The difference between fractions and decimals is merely one of notation and the use of methods appropriate to the notation, and we shall discuss them side by side. The relative importance of the two systems has changed considerably during the past twenty-five years or so.

Historically, fractions preceded decimals by some 3500 years. One of the earliest known mathematical writings, the Rhind papyrus, written about 1650 BC and itself a copy of a work written some two hundred years earlier, deals at length with problems involving fractions. The decimal point, however, did not make its appearance until the sixteenth or seventeenth century. The military engineer and mathematician Stevin, born in Bruges in 1548, wrote a work which brought widespread acceptance of the decimal system, although it is not certain that he invented it.

Fractions were introduced and developed as the need for them arose. By the time of the Romans, their use in measures of weight and length, where whole numbers of units proved insufficiently precise, made their manipulation of practical importance. Decimals were introduced as an invented system to provide an easier form of fractional numeration. It is the usefulness of both fractions and decimals that is the main reason for their inclusion in the school curriculum, although there is an elegance of structure in both systems that can appeal to intelligent children. It is, indeed, remarkable that eleven symbols, ten digits supplemented by a stroke or a point, should be able to express any quantity, however large or small, to any required degree of accuracy.

The use of vulgar fractions is becoming increasingly marginal in practice. Of course we shall go on using halves, quarters and the like: these are concepts that came long before the symbolization that made their study a topic in arithmetic. But there is, in effect, a generation gap between the older worker who would measure a length as 1 foot 2⅟₁₆ inches and the designer who would write 376 millimetres. We shall consider later the demands made on the skills of school-leavers and young workers in handling both fractions and decimals, but it is clear that extensive computation with fractions is, in general, obsolete in industry and commerce.

Parts and wholes

It is essential that pupils see fractions and decimals as useful descriptions of the world around them and not as topics of abstract mathematical number systems in which one has to do 'sums'. We all see and hear the words and may recognize some of the symbols, long before we meet them in school mathematics. Many children read road signs such as

Grantham 4¾

(written, incidentally, without the usual stroke, a circumstance that few people even notice). They see prices in a shop as £1.42, or tins of paint marked ½ litre. Very early on, a child will ask for half a slice of bread or be told to wait a quarter of an hour.

At this early stage, the terms used are merely names to the child; their meaning is confined to the context in which they are used. He knows that 28 is the number of his house, and he knows that he will not have to wait long if tea is in a quarter of an hour. His understanding of this last is unlikely to extend to the relation between the waiting time and a full hour. Again, the recognition of £1.42 on a price ticket certainly does not imply knowledge of the *relation* between 42p and £1, only that the price is more than £1 but less than £2. Gradually the child's knowledge will broaden until he grasps it firmly in a form adequate for his adult needs.

The word 'fraction' comes from the Latin *frangere*, to break. From the same root come 'fracture', 'fragile', 'fragment'. The basic concept is of a part broken, cut off, or removed, and given quantitative expression by the formalism of arithmetic. When we say that we have three-quarters of something, we think of a whole having, in some sense, been divided into four equal parts of which we take three.

There is, however, a danger in specifically teaching that a fraction is part of a whole; it is not clear from this description that it is the *comparison* between the part and the whole with which we are concerned. It must also be clear that the whole has to be divided, or thought of as divided into *equal* parts. There is also a later stage at which we want a fraction such as $\frac{7}{4}$, which can hardly describe a part in relation to a whole.

An adult usually has no difficulty with the semantics of the word 'fraction', but in teaching the range of the word is often contracted. The stock joke of the average family with $2\frac{1}{2}$ children needs to be taken seriously for our purpose. In using a fraction for comparison we need *two* quantities or measures to compare, one to be taken as the whole (or, more generally, as the unit), and the other which in relation to it gives rise to the fraction. Both measures are necessary. Half a piece of string is still a piece of string: it is only when we consider two pieces of string and say that one is half the length of the other, or that each is half of a piece now cut up, that the fraction one-half means anything. We cannot judge a length as half a metre unless we know what a metre is. The whole or unit is not always stated, but is always implied: 'half past three' means 'half of one hour after three o'clock'. A ticket marked '25% off' means a discount of $\frac{25}{100}$ of the original cost. In the child's first meeting with fractions, the 'whole' needs to be made explicit: he needs to see and handle the things being compared. The child needs the practical experience of filling bottles half full, of folding or cutting strips of paper into quarters, eighths, or even thirds. This last is a more difficult task, and it is a good language exercise for an adult to explain clearly why this is so.

The coloured rods used for number work form an excellent model of a system of fractions, and bring out the relation between the fraction and the earlier concept of division as the inverse of multiplication.

The length of the longer rod is four times that of the shorter, and in saying this we are expressing the longer in terms of the other. We could also say that the shorter rod is one-quarter the length of the longer. The two statements are equivalent, and should be recognized as such: both statements make a numerical comparison between the two rods.

One can also show the connection between fractions and the division process by considering an arrow diagram.

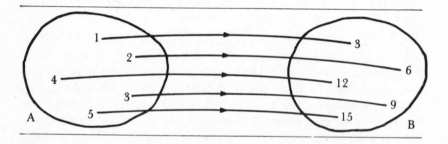

The arrow links members of set A to members of B by the relation 'multiplied by 3 is'; if the direction of the arrow is reversed it now reads 'divided by 3 is'. This is replaced by the relation 'is the third of'. Here we extend the original concept based on measures by taking the more abstract case of a relation between numbers.

One role of fractions such as a half, a quarter, a tenth or a hundredth, then, is to describe a relation between two quantities or numbers in which the smaller is compared with the larger considered as the unit. Taken the other way round, one expresses the larger in units of the smaller

> 1 centimetre is a hundredth of a metre
>
> 1 metre is 100 centimetres

Sometimes the comparison is made between a part and the whole which contains it, as when we speak of a bottle as half full, relating the liquid contents to the total capacity. But often, as with the lengths of two rods, they are separate objects and it is visually obvious which two lengths are being compared.

This distinction often leads to conceptual ambiguity.

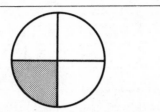

In the diagram the shaded part is one-quarter of the entire circle, but if we relate shaded and unshaded areas, the fraction required is one-third: the shaded part is one-third of the unshaded. The distinction is clear in this simple example, but both children and adults become confused over percentage increments and decrements, where the same principle is involved. The point needs to be discussed and made explicit in the early stages to save difficulties in the later.

We recommend that all simple fractions should be taught to children as words, as adverbs in expressions like 'half full' or as nouns with an indefinite article as in a half, a quarter, an eighth, a third, and a tenth. Until these are familiar, few other fractions should be introduced. So often later difficulty arises in number work because schools try to go too quickly.

If those selected fractions are now written as 1 half, 1 quarter, and so on, just as we write 1 apple, 1 book, then the ideas can be developed to include such fractions as 2 thirds, 3 quarters, and so on. The ideas become familiar *before* the notation is introduced. The concept now passes through three distinct stages, which we can illustrate by considering a rod which is 2 fifths the length of another. The three diagrams are shown at the top of page 78.

The first establishes the meaning of 1 fifth, the second makes a count of 2 fifths, the third shows a single rod equal in length to the 2 fifths. Coloured number rods do this job better than any other form of apparatus. The situation needs plenty of practice with the rods, and extension to shaded areas on squared paper or sectors of circles. Note that the basic process is still one of comparison: we cannot speak of a rod as representing 2 fifths unless we know what is taken as the unit.

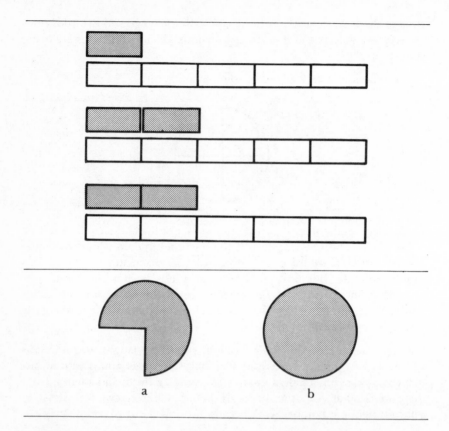

a   b

Sometimes it is difficult to avoid jumping to conclusions: one takes (a) as representing three-quarters on the tacit assumption that the unit would be (b). The example should be discussed with children, but not set as a catch question.

It is obvious that some pupils will themselves extend the concept to less familiar (and less useful) fractions, and that some will acquire the notation for themselves. There is no harm in this, but, for the sake of the greatest numeracy among the greatest number, actual teaching should not proceed until a child is completely at home with the concepts discussed. Criteria for success are worth listing. The list shows how extensive can be a child's grasp of a concept, without its having been linked to formal arithmetical notation and processes of calculation. A child is not ready to proceed with formal work in either decimals or fractions (and by implication no adult can begin to be numerate) until able to:

1 Identify examples of fractions in everyday life.

2 Identify examples of decimals. (This does not imply understanding of the notation.)

3 Read simple examples expressed in words.

4 Using simple fractions such as 1 half, 1 quarter in given practical examples.

5 Make up such examples, using coloured rods, pieces of card, or other material.

6 Form situations that illustrate fractions such as 3 quarters, 2 thirds.

7 Decide by making up situations with rods or other apparatus which of two simple fractions of a unit is larger.

8 Experience, identify, and form a few examples of the equivalence of two fractions (e.g., that 1 half is equivalent to 2 quarters).

9 Estimate fractions in simple examples (e.g., a bottle half full).

10 State or show between which of two consecutive whole numbers a given decimal lies (e.g., £12.35 lies between £12 and £13).

This is an impressive list of skills, and we emphasize that they are all prior to formal work on decimals or fractions. We should also expect children to use and hear fraction words freely in context, without necessarily being aware of their relational significance. We all do this frequently whether numerate or not. The teacher will say, 'I want half of you on this side of the hall and the other half over there', or 'Bring me the half-litre measure'. We may exclaim '£4.85! That's nearly £5 for something not worth buying!' Such usage can be made the starting point of one more stage in the number concept.

Extension to ratios less simple than those discussed will of course be made, but we recommend that the first stages in formalizing the work in fractions should come before this extension. The concept of angle as a fraction of a complete turn requires understanding of such fractions as $\frac{43}{360}$, but it seems reasonable to return to such examples only after the notation is fully understood. If the notation $\frac{1}{2}$ for a half, $\frac{1}{4}$ for a quarter, even $\frac{3}{4}$ for 3 quarters is used when appropriate without comment, these compound forms will be learnt as single ideograms, as we suggested earlier for 10, 11, 12. Many children will get them from road signs, and many parents would take the opportunity of discussing such examples with them.

Fractions as measure

If we say a rod is 2 metres long, we are using the number 2 to record a measure, the count of the units we put end to end to equal the length of the rod. If we say the rod is 2½ metres long, we are no longer counting units. In the counting process we are dealing with discrete objects, and we now wish to deal with length, which is a continuous measure. It is clear that a fixed unit such as a metre is not necessarily stepped out exactly against any given length, and fractions, either vulgar or decimal fractions, become one possible way to meet the difficulty. It should be realized, however, that the use of a fraction of a unit to deal with the excess of a length over a whole number of units, as with the rod whose length is 2½ metres, is not the only way. An alternative is to have available a wide range of units from which we choose one of convenient size for the job in hand.

This, of course, is what happened in the history of measure: units were devised to meet local or special needs so that at most only simple fractions would be called for. This process left us with a multiplicity of measures of the same kind. For length, for instance, we had leagues, miles, furlongs, chains, fathoms, yards, feet, hands, inches, points ... Some of these interesting historical survivals will remain in informal use for some time, but we no longer need to calculate with them except in special circumstances. They are being generally replaced with the Standard International (SI) system of units based on the original metric system. This makes very modest demands on number skill, since there is only one standard unit for each kind of quantity to be measured with easy multiples and submultiples of the unit to give the necessary range.

We are discussing numeracy as a background to the *use* of measurement rather than its techniques. We only want to make the point that by using a sufficiently small unit fractions may be avoided altogether. This in fact is recommended in engineering or building practice for detail constructional dimensions. An opening to be fitted with a door, for instance, would be measured as 776 millimetres by 1953 millimetres, using the millimetre, a unit sufficiently small not to need subdivision for the purpose in hand.

It is nevertheless true that a consideration of measure is the most effective way to introduce the notation for decimals, and we shall use it for this purpose, together with the established notation for the currency, in the next section.

Summary

This chapter has been devoted to an informal development of the fraction concept, omitting any detailed account of the notation. Above the dotted line in the flow diagram is the conversational use of fraction words from which we began:

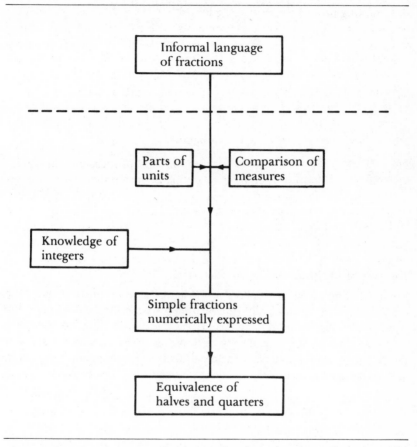

At this stage the concept is verbal rather than numerical, but the pupil's knowledge of number is being used to express the fractions as symbols.

CHAPTER 6

THE DECIMAL NOTATION

The use of decimals

Children of today, unlike those of an earlier generation who may now find calculation with decimals difficult, will at least meet the notation in use from an early age in expressing the currency, although they may not easily associate the two. They are also likely to become familiar with denary sub-units. The odometer of a car registers in tenths of a mile, not in furlongs, and petrol pumps record tenths of a gallon, not quarts and pints.

Apart from the general use of halves, quarters, and eighths in cir-cumstances that do not require extensive calculation, any application of more difficult fractions is residual or specialized. The applications of decimals are wide and of increasing importance, and it seems will go on in-creasing as the technical uses of vulgar fractions become less. A numerate person today must be able to handle decimals in expressing measures and interpret data given in decimal form. Computation with decimals should at least include addition and subtraction, and the multiplication and division of decimals by one-digit integers.

At this point the borderline between the essential and the desirable becomes rather vague. One can live a long time (out of school) without ever needing to multiply or divide two decimal fractions, one by the other. One feels, however, that a numerate person should be sufficiently aware of the properties of the notation to realize how, for example, the product 0.9 × 0.7 is related to 9 × 7, or how 6 ÷ 0.3 is related to 6 ÷ 3; so one should at least try to get all children familiar with examples of this kind. A good general principle is that computation should not go beyond the number of significant figures that are likely to arise in day-by-day measurement or recording.

It is the knowledge and number skill thus demanded that must be met by the schools. Here it is likely that they must make a break with tradition and formalize the teaching of the decimal notation before vulgar fractions are taken beyond the early stages listed at the end of Chapter 5.

The tradition is a long-standing one that goes back to its origins in eighteenth-century commercial and nineteenth-century technical arithmetic. Work with vulgar fractions has been done for so long that it is done without question; but there are now no practical reasons why it should continue as it has. There are also very good theoretical reasons why decimals should be introduced earlier, because they enable us to deal with parts and with subunits merely as an extension of the notation. That this was the intention of the early writers such as Stevin is quite clear, but what happened when arithmetic became a topic for general education was that decimals were put in as an additional and more sophisticated topic, as logarithms were at a later date.

We have already recommended, as a useful measure in counteracting in-numeracy, a more careful approach to the notation, and when this has been done, work in decimals follows much more easily. We now suggest that schools should reverse their tradition, and introduce the Four Rules with decimals before beginning formal work with fractions. We recom-mend that the notation should be studied, and very simple computation begun, so that the pupils can deal with easy problems involving measure, before approaching a further study of fractions, which would begin with the numerator/denominator and the key concept of equivalence.

It is perhaps worth recalling the transition period that was planned to follow the United Kingdom's change to decimal currency. There were quite detailed arrangements for double pricing and for shopkeepers who so wished to continue transactions in the old currency for twelve months. In the outcome, few shops took advantage of the arrangement, and those that did soon abandoned it. There was, also, in response to public outcry, an agreement by the government to leave the old 6d. coin in circulation. Within a few months, the new currency had become fully absorbed except among those who consciously resisted it, and within very few more the sixpence had practically disappeared. Decimal coinage proved to be simple for most of us, although it is true that elderly people sometimes had more difficulty in changing the habits of a lifetime. It is remarkable, and a matter of psychology rather than logic, that most people do not associate our currency with general arithmetical processes, and hence do not see it as an application of decimal fractions. They learn to work with the currency, but this does not seem to help with calculations involving decimals as pure number processes. American children who grew up with dollars and cents find decimals no easier than English pupils. The solution to the problem is a complete reordering of the work in vulgar fractions and decimals, using the currency and SI measure as a lead-in to decimal concepts.

Extension of place value

There is much scope for alternative approaches to the teaching of place value, particular for decimal fractions. We shall outline one approach to make clear the ground that ought to be covered, the range of understanding that we want everyone to possess.

One can begin, with only passing reference to the place-value notation as such, by making explicit the function of the point in the notation for the currency. We count in pence from 1p to 99p, and then at 100p we change to £1. We could write 200p, but we usually do not; we write £2. If there are both pounds and pence, we write both, separating them by a point. This point is *merely* a separator, and the banks ask us to use a dash instead. In keeping accounts we use ruled columns for £ and p and do not use the point at all.

The child who is, as he should be, familiar with the essential column line-up of digits will soon see exactly why £1.1 does not do for 1 pound and 1 penny. Since we need to go up to 99p before reaching the £1 we shall have quantities such as £1.23, where 3 is the units digit and 2 the tens for the amount in *pence*. An odd penny must be added to the 3 units, not the 2 tens, and so we would have to write

£1.23
£1. 1

This leaves an empty column in which one inserts the zero invented for this purpose, getting the correct forms

£1.23
£1.01

If we have 10p, there is a zero already written, so 1 pound and 10 pence is

£1.10

The discussion will formalize for the child what is probably already known perfectly well in usage. It deals with the special unit 1p of which 100 make up £1, and does not yet introduce the decimal fraction proper. It nevertheless prepares the ground.

The next step is best taken using measures of length. Measure such as one-and-a-half metres will be familiar, estimated by using ungraduated metre rules. From here, using rules graduated in tenths, preferably in alternative coloured sections, the transition to a notation for tenths can be made. The word decimetre can be used at the teacher's discretion, and we would in fact recommend it in the interim. We strongly recommend a set of tapes for children to measure, labelled A, B, C . . . or numbered for identification and previously cut to whole numbers of decimetres. The child will, for example, have a tape whose length is 1 metre 7 decimetres. Whether he measures this using an ungraduated metre rule together with the divided one, or whether he is shown how to step out the length of the rule along the tape is a matter for the teacher. What matters is that he records a length of 1 metre and 7 decimetres. We ask the child to record this as 1.7 metres. It will eventually become clear after discussion that the function of the point is to show which digit is the unit. To the left of it lie the units and tens, to the right the tenths. Any other indicator would do, and in fact the odometer of a car usually uses digits of a different colour for tenths, but from now on we agree to use the point, and call it the decimal point. It is not necessary to use the word decimetre from now on.

It remains for the pupil to measure a collection of objects chosen as appropriate, at first with prepared tapes or laths which can guarantee that what is wanted is reasonably clear and unambiguous. The decimetre rods included in sets of structured number material are also of use. Lengths less than 1 metre can be introduced, but only after discussing why it is necessary to begin the record with 0 when there are no whole units but only tenths. Thus the child collects and records measures such as 2.3 metres, 1.7 metres, 0.6 metres.

Computation with decimals

Use of marked rods and the structured arithmetic apparatus should soon establish the equivalence, if it needs establishing, of 10 tenths of a metre and one whole metre; one can then take the basic step of column addition.

Using an example

0.6 m + 0.7 m = 1.3 m

This can be set down in the familiar form

$$\begin{array}{r} 0.6\,m \\ +0.7\,m \\ \hline 1.3\,m \end{array}$$

Teachers will note that it does not help to write a carrying figure here, which would merely slow up the single mental stage. We write directly

$$\begin{array}{r} 6 \\ +7 \\ \hline 13 \end{array}$$

and so similarly

$$\begin{array}{r} 0.6 \\ +0.7 \\ \hline 1.3 \end{array}$$

The zeros simply keep the columns clear and are ignored in adding. If this is done thoroughly and without attempts to hasten understanding, the use of the decimal notation is now, almost without the pupil's realizing what has been achieved, complete in its first stages, the loss of which is so often the cause of later puzzlement.

The child will soon agree that in working with decimals it is better to use a zero to keep column alignment even if there are no tenths, so that one has column entries such as

2.3
4.0
3.6
0.1

The points must be kept ranged one under the other so that units and tenths are never confused. Once this is clear, the pupil will go a long way before the next difficulty is met, providing his work is planned to this end.

All four rules can now be dealt with, and by choosing multipliers less than 10 and small divisors that leave no remainder, the pupil is launched into decimal arithmetic, through carefully planned activities in measuring, recording, and eventually calculating using pure number.

Examples of calculations would be

$$
\begin{array}{llll}
0.7 & 6.2 & 4.7 & 2\overline{)7.6} \\
+5.8 & -2.8 & \times 3 & 3.8 \\
\hline
6.5 & 3.4 & 14.1
\end{array}
$$

We shall not discuss these examples, since they would be taught by simple extension from the methods recommended in Chapter 4. They can, with advantage, be demonstrated with coloured rods using the decimetre rods as units, or the work can be done on an abacus with a column separator or coloured section, as in the diagram.

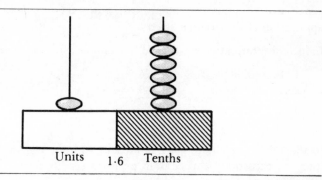

Units   1·6   Tenths

Note the position of the multiplier 3 in the example of multiplication above. The right answer may well be obtained wherever the 3 is put, but our ultimate aim is to reduce the failure rate for the product 4.7 × 0.3. The school will use column headings units and tenths at its discretion.

The pupil should also be explicitly aware of the decimal equivalence

1 tenth of 1 metre is 1 decimetre

10 decimetres are one metre

The next stage is to extend the notation to include hundredths, after which the pattern should be established and can be extended indefinitely to the right as far as is needed. The hundredths should be done practically. Suitably marked metre rods, calibrated in decimetres and centimetres (*not* millimetres) present both a model of the system and a practical measuring instrument. The needed equivalences should be readily seen

1 hundredth of 1 metre is 1 centimetre

100 centimetres are 1 metre

10 centimetres are 1 decimetre

so a measure given provisionally in words as

1 metre 3 decimetres 5 centimetres

becomes 1 metre 3 tenths 5 hundredths

and we write

1.35 m

We can think of this equally well as 1 metre and 35 hundredths, so that we link up at once with the known notation for 1 pound 35 pence, there being 100 pence to a pound just as there are 100 centimetres to a metre. In effect, the notation for money, taught *ad hoc* on the basis of the child's probable experience of shopping and prices, now takes its place as one more application of the decimal system.

Calculation can now be extended to examples such as these:

$$
\begin{array}{llll}
1.36 & 9.36 & 4.29 & 2)\overline{5.26} \\
+2.07 & -6.18 & \times 2 & \overline{2.63} \\
\hline
3.43 & 3.18 & 8.58 &
\end{array}
$$

Readers will accept that, given familiarity with three-digit operations with integers, classroom work should progress readily from the examples on page 87 to the above.

Whether the development need continue to three or more places of decimals is a question of understanding rather than use. In practice 1.237 metres, if needed, is usually written 1237 millimetres, and is not really everyday usage for the man in the street. But dealing with one more digit will certainly help fix the pattern of the notation, and fortunately such numbers can be shown in action on a metre rule, graduated, as they usually are when supplied for work in school science laboratories, in centimetres and millimetres. There are 1000 millimetres in 1 metre and hence the fraction is 1 thousandth. The notation follows easily from what has gone before. A modern textbook or technical work would never write

1 m 2 dm 3 cm 7 mm

but 1.237 m or 1237 mm

Since the object is to help with the notation rather than to deal with the techniques and conventions of measure, one would not at this stage mention the practical advantages of recording 1237 millimetres. Children should examine metre rules with millimetre graduations. They may also use the rules for measure, but one does not expect yet to attain three-figure accuracy in measurement. Rules supplied for actual classroom measure for younger children should be in decimetres or centimetres. The same numerical relation holds for litres and millilitres, but a litre measure is not usually graduated in millilitres.

The pupil has now, by very simple stages from the acceptance of a suggested notation for tenths, reached the point where, in number work, he can tackle calculations such as

4.236 + 3.217

6.932 − 2.816

6 × 5.203

6.422 ÷ 2

That is, given that a child can handle integers within the limits likely to be met in everyday life, there is very little extra effort needed on the part of teacher or child to extend this work to decimal fractions, again within the limits beyond which one verges on specialism.

It should be noted that whereas the decimal system we have just discussed arises naturally and easily from the denary-based notation for whole numbers, there is no such natural extension for the traditional notation for vulgar fractions. A 3 followed by a 2, as 32, is read thirty-two, 3 tens and 2 units, but followed by ¾ is read as 3 units and ¾ of a unit. Those acquainted with algebra will recognize a similar difficulty arising later, when the notation 3n has to be understood as 3 lots of n. Fortunately, a knowledge of algebra is hardly required for general social numeracy.

Decimals as number

Although we suggested an approach to the decimal notation through the use of measures of length, decimals form a complete abstract number system. The mathematics teacher is probably more interested in this aspect of the topic than society in general is likely to be. Decimals help do a job, and if one can use them one can do the job. An understanding of their nature as numerical entities seems to go beyond the basic and essential, but we hope that many intelligent adults would, at some stage in their education, have discussed the matter if only as an exercise in logical thinking.

One needs both to discriminate and to generalize. We arrived originally at the concept of number by counting discrete objects, but the decimal fraction clearly goes beyond this. One can count 6 metre rods laid end to end, but one cannot count 6.237 of them. What we do, of course, is to set up an imagined subunit, in this example a thousandth of a rod, and count that, using the numbered graduations on the rod to keep a tally. This subdivision can be continued indefinitely, in imagination if not in practice, and in this way we bring the decimal into the system, with a notation that includes denary submultiples as well as multiples of ten.

What is much more difficult to grasp is that vulgar fractions and decimals are simply different ways of writing the same kind of number, the kind that is arrived at by subdividing a unit into an agreed number of parts, followed by counting out the subdivisions. Writing 0.3 as $\frac{3}{10}$ makes this easier to see, but only a knowledge of the notation explains why the former is so much more convenient. If decimals had been invented first, it is extremely unlikely that vulgar fractions would have gained acceptance as an elementary study, although they would be needed later for dealing with higher order concepts such as ratio.

Halves, quarters, thirds, eighths would certainly be used and would probably have been given an mathematical expression that would enable them to be combined arithmetically if needed, but examples such as

$$\frac{\frac{2}{3} \text{ of } \frac{3}{4} - \frac{7}{8} \div 1\frac{3}{4} + 1}{5\frac{1}{2} - 2\frac{1}{4} \div 1\frac{1}{4} \times 1\frac{1}{2}}$$

would be included with such feats as trying to do long division using Roman numerals. The example exactly as printed, and with no help from brackets, came from a book of commercial arithmetic for students of office routine printed in 1965, and shows how wide is the gap between textbooks and practical activity, and how repellent number work must be to those who do not have the knack – the very superficial knack – of 'handling figures'.

Moreover, such examples were, no further in the past than the 1965 office arithmetic, a hurdle that had to be cleared before decimals were begun. Our suggestion that decimals should be begun first, and our claim that much later innumeracy would be avoided if it were, is not likely to produce a sudden change in school curricula. The older traditions, with no basis other than historical accidents, are firmly established in the self-perpetuating sequence of pupil-teacher-pupil. But the matter is so serious, and the failure of current programmes so much a matter of concern, that a general discussion of the approach to number work is necessary. Our claim needs careful examination.

Multiplication and division by ten

The operation of multiplication by 10 was presented in Chapter 4 as the key process for multipliers of more than one digit. It is, similarly, the key step for an understanding of decimal computation, and unless properly presented for integers the transition is likely to cause trouble. If it has been properly presented, there should be no difficulty: indeed, if it has been thoroughly understood as a process with integers, the next step is the obvious one.

Teachers will have their own methods for dealing with the topic, but it should be clear what happens. When 73, say, is multiplied by 10, the 3 units become 3 tens, the 7 tens become 7 hundreds, and so all digits shift one place to the left, leaving an empty column which is then marked with 0.

$$\begin{array}{r} 73 \\ \times 10 \\ \hline 730 \end{array}$$

We have already discussed the spurious rule about adding zero, which misleads in examples such as

$$10 \times 73 = 730$$

The requirements of column alignment are not obvious when products are set out in this way. Children taught such a rule are very apt to write

$$10 \times 7.3 = 7.30$$

Given familiarity with the notation, the actual stages are clear. A tenth taken 10 times is one unit, and 3 tenths taken 10 times are 3 units

$$10 \times 0.3 = 3.0$$

The final zero is inserted in this example to draw attention to what has actually happened: the tenths digit has moved along to the units column. So we have

$$\begin{array}{r} 7.3 \\ \times 10 \\ \hline 73.0 \end{array}$$

The correct rule already taught does not change, only the setting. To multiply a number by 10, move all digits one place to the left. If a number happens to include a decimal point, whose job it is to mark which digit is the unit and where the denary subunits begin, this makes no difference at all. The decimal point stays where it is.

Children who have been taught the spurious 'zero rule' are often taught another rule *ad hoc*, to cope with decimals.

Second spurious rule: to multiply by 10, shift the decimal point one place to the right.

This again gives the right answer for the wrong reason, and convinces children that decimals are different entities from ordinary numbers, having rules of their own that have to be remembered. It is difficult to justify the existence of such an approach: the only excuse is that arithmetic teaching was once done by people who had themselves not learned the essential structures of number, so that the faulty logic has been repeated from one generation of pupils to the next. Older textbooks are full of such rules, often elaborated to a point where they are difficult to follow even as instructions taken on trust.

It is now a straightforward matter to extend the process to hundredths and thousandths, either directly or by appeal to the markings on a metre/millimetre rule, and we have such results as

$$
\begin{array}{r}
2.632 \\
\times 10 \\
\hline
26.320 \\
\hline
\end{array}
$$

Note that in writing the result we have filled in 0 in the vacated column of thousandths. This is probably the best thing to do, but when such a result is being transferred elsewhere one would not bother to use the zero, and the child should see why. Later, in recording measures, the pupil sees that there is a use for such zeros.

If this section is read in parallel with the section in Chapter 4 it will be seen that there is no conceptual break. There is an equally smooth transition for the process of division by 10. The decimal point maintains his station as the marker, and the troops of digits all go one place this time to the right as hundreds become tens, tens become units, units becomes tenths, tenths become hundredths . . .

Correct rule: to divide a number by 10, shift all digits one place to the right.

Since we only use the point when there is a need for it, and since the units of an integer become tenths on division by 10, one often introduces a decimal point not previously written

$$23 \div 10 = 2.3$$

or

$$
\begin{array}{r}
10)\overline{23} \\
\hline
2.3
\end{array}
$$

Some teachers like to set the work out in full

        2.3
10)23
    20
    ——
    3.0

Some even introduce the decimal point into the dividend to show that one is now thinking in terms of a tenths column, at first not required but now called for by the splitting up of the 3 units into tenths. One might then see

        2.3
10)23.0
    20.0
    ————
    3.0
    3.0
    ————

All these forms are correct: they all get the right answer at various levels of contraction. Any difficulty lies in the sudden appearance of the point. The number 23 is not normally written with a decimal point: one does not need to separate units from nonexistent tenths. But the tenths come into existence as the 3 units in 23 are divided into ten parts. Written in columns there is no confusion,

10(23
    ——
    2.3

An abacus, suitably provided with a coloured portion or a marker to label a wire on which each bead represents a tenth, also makes it quite clear

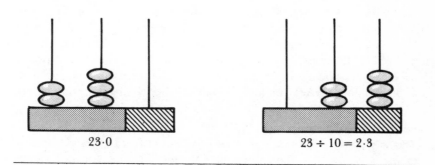

23·0   23 ÷ 10 = 2·3

Writing the result in one line gives $23 \div 10 = 2.3$, and this has given rise to one more misconception, often quoted as a process.

Third spurious rule: to divide a number by 10 insert a decimal point.

We give the three spurious rules as a calculated risk. Teachers are well aware of the danger of negative instructions given to children, but by quoting them to adult readers we hope to keep them out of the classroom. Once in, they establish themselves with the status of revealed wisdom.

It should also be clear that the exact column alignment called for in Chapter 4 is even more important for division. One could draw a vertical line on the page to mark the beginning of the tenths, but the convention is to use the decimal point, and those must be kept ranged as if on a line.

By this time pupils should be becoming reasonably familiar with the notation, and move easily to multiplication and division by 100. Division of a single digit by 100 shifts it 2 places to the right, into columns that are not already marked because they were not needed. We then have to introduce a place holder zero into the tenths column, and recommend a further zero in the original units column to ensure correct alignment

$6 \div 100 = 0.06$

The alignment can easily go awry in columns of figures if this is not carefully done

```
 12.3
  0.03
 30.1
  1.01
127.0
```

Reference back to page 93 and the example $23 \div 10 = 2.3$ shows that we have here, incidentally and without explicit mention of it, dealt with the 'remainder', which would have appeared as 3 in the earlier work. There is no need to elaborate at this stage, since the notation has done our work for us! Ten does divide exactly into 23 if we allow the use of our extended decimal number system, because it divides the remaining units into tenths.

Further computation with decimals

The way ahead from this thorough preparation is now straightforward, taking its own line at the discretion of the teacher. The sequencing to be considered in preparing classroom work can be shown by the examples that follow, which also suggest the limits one should apply to the number of digits involved in such calculation. As often in mathematics, the teacher can take the child a long way by selecting examples which avoid more difficult stages in understanding. As a general rule, the simpler the example the more likely it is to be met in practical life. So many adults have been deterred by complicated calculations that have no use outside the classroom. Long calculations are often justified by the claim that some children like them. We feel that such children should move on to more sophisticated concepts rather than spend their time practising mechanical skills.

This is a possible sequence of examples that could be considered for multiplication

1   3×2.63
2   10×2.63
3   30×2.63
4   32×2.63

Multiplication by 30 is done in one step which can be analysed as two: multiplication by 10 and then by 3 (or the other way), so that the digits are shifted one place to the left. The zero is left in the example

$$
\begin{array}{r}
2.63 \\
\times 30 \\
\hline
78.90
\end{array}
$$

Multiplication by 32 now follows as a combination of types (1) and (3) above, as in Chapter 4,

$$
\begin{array}{r}
2.63 \\
\times 32 \\
\hline
78.90 \\
5.26 \\
\hline
84.16
\end{array}
$$

For division one would consider the types

1   $2.60 \div 4$
2   $73.26 \div 20$

Setting out both calculations in full, with all subsidiary zeros inserted, one gets

```
    0.65            3.65
4)2.60        20)73.26
  2.40           60.00
  0.20           13.00
  0.20           12.00
  0.00            1.00
                  1.00
                  0.00
```

These will undoubtedly be contracted or modified to suit the needs of the teacher. Note that using a two-digit divisor we have chosen a multiple of ten and a dividend that leaves no remainder. This allows the products 60, 12, and 1 that are to be subtracted to be determined mentally, with the digits in the quotient obtained as the inverses. One says: '3 lots of twenty are sixty, put down 3'.

The range of possible decimal calculations is now quite large. The learner is still very vulnerable to the Uncle Pumblechooks of this world, who on hearing that the child is doing decimal division will want him to calculate $0.023 \div 0.736$, but this is one of the hazards of enlightened and planned education.

The range of skills in manipulating decimals which are to be the objectives of teaching number work so far can be listed

1 Multiply and divide decimals by integers of up to two digits, without remainder.

2 Divide decimals by one-digit integers, with remainder terminating after one digit.

3 Multiply integers by one decimal digit.

4 Divide decimals by multiples of 10.

5 Recognize and make up practical situations requiring the above operations.

Of these, the fifth is the most important for full numeracy. If it proves difficult for the teacher or the examiner himself to find a wealth of plausible examples, then it follows that the work is going beyond what a basic course requires. If there is a general use for $0.023 \div 0.736$, then full general numeracy requires the ability to tackle it. If one can only find a use by searching diligently, then such a calculation becomes appropriate only to an extended numeracy. This does not mean that it is omitted from the curriculum, only that it is not considered as part of a basic course. The question of understanding is crucial. This is why we try to introduce each new operation through an activity involving concrete objects, and keep the actual numerical examples simple until the principles have emerged. This passage from concrete activity to abstract understanding is not a matter of education but of maturation, and it is not possible to speed it up. There is always the danger, as we have seen with the three spurious rules, that children will acquire early on a 'rule' which produces the correct answer in a given case but which grossly misleads in a later situation. The skills with which we are concerned lend themselves to easy description in annotated algorithms, and it is easy to be misled by a child's acquisition of a skill into thinking that the doing is accompanied by understanding. If he is not confidently numerate, he will not be able to recognize an unfamiliar situation requiring a skill he apparently possesses, nor will he be able to retain it without constant mechanical drill.

The product of integers with decimal multipliers is best dealt with, for our purpose, by a tacit appeal to the commutative law. In logic, multiplication having been described as repeated addition using the phrase 'lots of', as in 4 × 0.3 which is 'four lots of 0.3', one cannot pass without comment to '0.3 lots of'. Since pupils should be quite used to taking products in any order, it is only necessary to point out that 0.3 × 4 can be done by thinking of it as 4 × 0.3

The argument can, of course, be made more convincing, if, but only if, the pupils are able to follow explanations that require more than a couple of stages. A possible set of stages would be

2 × 4 is 2 lots of four, or 8

1 × 4 is 1 lot of four, or 4

0.1 × 1 is 1 tenth of a unit, or 0.1

0.1 × 4 is 1 tenth of 1 lot of 4, or 0.4

0.3 × 4 is 3 tenths of 1 lot of 4, or 1.2

It is a great help if the learner can see the formal equivalence of multiplication by 0.1 and division by 10. Informally 0.1 × and ÷ 10 are both verbally expressed as '1 tenth of'. Since division by 10 shifts all digits a place to the right, so does multiplication by 0.1. Using a multiplicand (the number multiplied) of three digits we get

$$\begin{array}{r} 526.0 \\ \times 0.1 \\ \hline 52.6 \end{array}$$

Each unit becomes a tenth, each ten a unit, and so on. This kind of discussion is more difficult to follow when expressed in writing than it is when run through informally over a period of time as occasion offers: the schoolchild should not be expected to grasp it the first time in so compressed a form. In the early days, pupils could well begin by inserting a decimal point after the last digit, to make sure that the multiplier is put in the right place.

We do not recommend an attempt to teach the multiplication of a decimal fraction by a decimal fraction at this stage. The process will be dealt with in a later chapter, where we shall argue that it probably goes beyond what is a basic requirement, at least in the form traditionally given.

Summary

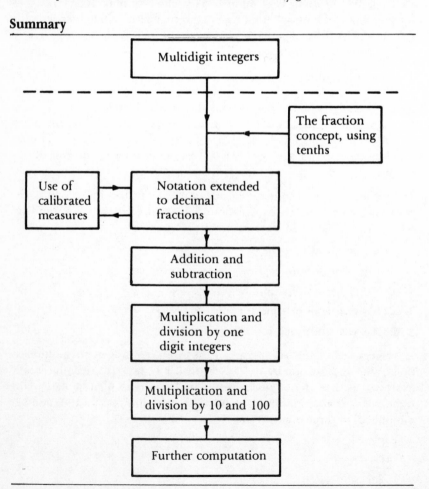

The topics recommended at this stage of learning are given as a flow diagram. Note that the extension of the denary notation to include decimals is possible because the fraction concept can now be brought into use. The notation for vulgar fractions is not used here.

CHAPTER 7

OPERATIONS WITH FRACTIONS

Numerators and denominators

In Chapter 5 we suggested that children would probably learn, informally, the notational forms ½, ¼, ¾ through seeing them in use. Such fractions would continue in use throughout any work on the beginnings of the decimal notation. Further work in fractions should be introduced once the decimals are off to a good start.

By this time the forms ½, ¼, ¾, and the ideas they express will be more firmly in mind, and we can now build on them. This means that we should provide a variety of illustrations for each fraction considered. This is important for a more mature understanding of the fraction in use. For example, the shaded part of each of the following illustrates precisely ½ of the diagram taken as a whole

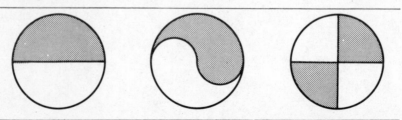

The third diagram is suggested because the shaded half appears as two separate quarters.

It is now time to draw attention to the meaning of the 'top' and 'bottom' of a fraction, a notation which expresses the 'whole and part' concept that is one of the starting points. The notation $\frac{3}{4}$ we describe as forming a fourth part or quarter and then taking 3 of them. One needs to emphasize this procedure to draw attention to the different roles of the digits above and below the horizontal, or before and after the oblique stroke as in 3/4. Questions such as 'what have these fractions in common – $\frac{3}{8}$, $\frac{4}{8}$, $\frac{1}{8}$, $\frac{7}{8}$, $\frac{2}{8}$?' encourage the answer that they are all eighths. This is why the number at the bottom is properly called the denominator because it names (Latin: *denominare*) the parts into which the whole is divided, but there is no need for the children to know the term at this stage. The number at the top, which tells us how many eighths we have, is similarly called the *numerator* because it tells us how many of the parts are taken. Children may well confuse these words but are unlikely to confuse 'top' and 'bottom' as used of fractions.

Later, one might expect these italicized words to be known, but in the meantime one should not add to the difficulty of using a not very obvious notation by using even less obvious words. Once the idea is grasped, the concept of fraction can easily be extended, as a concept, to any denominators. The meaning of $\frac{2}{13}$ or $\frac{24}{31}$ should be quite clear, although it should also be clear that such fractions are rarely used.

We do not, at this stage, talk of $\frac{3}{8}$ as 3 divided by 8. It is the *unit* which is divided into 8 parts, and 3 of these are taken.

$$\tfrac{3}{8} = 3 \times \tfrac{1}{8}$$

Nor should we, at least in this context, write $\frac{12}{4}$ when we mean 12 divided by 4. It is not wrong, and will be used freely at a later stage in using mathematics, but here it confuses the issue and detracts from the significance of the denominator in naming the sort of fraction we have. It is also advisable to work now and then with tenths, stressing that the fractional and the decimal notations here represent exactly the same quantities. Vulgar fractions and decimals are equivalent, and the adult should be able to use one or the other according to the convenience of the situation. Traditional teaching must largely take the blame if he does not.

Equivalent fractions

The concept of equivalence is basic to mathematics at every level. To say that two things are equivalent is, in effect, to say that they are interchangeable for the purpose in hand. It does not mean that they are identical for all purposes. The word 'six' is equivalent to 'half a dozen' if one is buying buns, but not if one is using a typewriter. If I cut a pork pie into four equal parts and eat all of them, I have eaten a whole pork pie; but a customer who asks for one in a shop is unlikely to accept the four segments.

If we are concerned with a total measure or numerical quantity as such, then, but only then, can we write

$$\tfrac{1}{2} = \tfrac{2}{4} = \tfrac{4}{8} = \tfrac{8}{16} = \tfrac{5}{10} = \tfrac{15}{30} = \ldots$$
$$\tfrac{2}{3} = \tfrac{4}{6} = \tfrac{8}{12} = \ldots$$

The above form two sets of equivalent fractions.

The pupil is often told that 'multiplying top and bottom' of a fraction does not change its value. This is one more example of a short cut that misses the point to be understood. We are actually changing the subdivision of the units as named by the denominator, from halves, say, to tenths. Each subdivision is now a fifth of what it was and hence we take five times as many of them, multiplying the numerator by 5.

It is worth noting that the equivalence

$$\tfrac{1}{2} = \tfrac{5}{10}$$

implies that we also have merely as a change of notation the equivalence

$$\tfrac{1}{2} = 0.5$$

One reason why quarters, eighths, and sixteenths are certain to continue in use as vulgar fractions is that they are produced by repeated halving. They are not so readily expressed as decimal fractions.

The importance of equivalent fractions is that their use provides a technique for adding and subtracting fractional quantities. Addition, as the total count obtained by putting together the counted sets of similar objects, as in

2 eggs and 3 eggs make 5 eggs,

is not applicable if the objects are dissimilar. Two eggs and 3 good wishes, however ingeniously we manipulate them, do not readily show number bonding in use. Nor does 1 quarter added to 3 eighths, although here we are at least discussing the same sorts of entities. If we use the equivalence

$$\tfrac{1}{4} = \tfrac{2}{8}$$

we can now write

$$\tfrac{1}{4} + \tfrac{3}{8} = \tfrac{2}{8} + \tfrac{3}{8} = \tfrac{5}{8}$$

In this way all fractions likely to be met can be added or subtracted as required.

If the total count given in the combined numerator is greater than the denominator we use equivalences such as

$$\tfrac{2}{2} = \tfrac{3}{3} = \tfrac{4}{4} = \tfrac{5}{5} = \ldots = 1$$

For example:

$$\tfrac{2}{3} + \tfrac{3}{4} = \tfrac{8}{12} + \tfrac{9}{12}$$
$$= \tfrac{17}{12}$$
$$= \tfrac{12}{12} + \tfrac{5}{12}$$
$$= 1\tfrac{5}{12}$$

Some teachers would prefer to omit the third line.

We take it that the work as it proceeds should be illustrated by many diagrams or exercises using structured material. Here is one example:

Two pies are cut into sixths. The shading shows that 2 thirds of a pie are equivalent to 4 sixths, and that 12 sixths are equivalent to 2.

It is worth recalling the stages by which addition of fractions was once taught. We first find a number called the Least Common Denominator or Multiple, usually abbreviated to LCM. This could be the simple product, but often examples were constructed with denominators such as 4 and 14 where it was not. The work was then set out with its appropriate liturgy:

$$\tfrac{3}{4} + \tfrac{5}{14} = \tfrac{21 + 10}{28}$$
$$= \tfrac{31}{28}$$
$$= 1\tfrac{3}{28}$$

'4 into 28, 7; 3 sevens are 21, put down the 21. Plus. 14 into 28, 2; 5 twos are 10, put down the 10. 21 + 10 = 31. Result 31 twenty-eighths equals 1 and 3 twenty eighths.'

Such examples were repeated until they became mechanical, and any initial explanation long forgotten.

(Readers who happen to be mathematicians at once recognize this as the general definition for the sum of two rational numbers, with the added difficulty of finding the LCM of their denominators instead of using the product as it appears in the general case:

$$\frac{a}{b} + \frac{c}{d} = \frac{ad + cb}{bd}$$

They will also agree that this is irrelevant to the problems of numeracy.)

As the relative importance of decimal calculations increases rapidly, and that of fractions decreases, it seems unreasonable to include, in any test for general numeracy, calculations in fractions for which a common denominator cannot be chosen at sight.

Multiplication by integers

If the function of numerator and denominator is understood, multiplication of a fraction by an integer should be straightforward, and children can be expected to grasp this stage easily. If we have 3 boxes, each of 6 eggs, we have 3 × 6 or 18 eggs. If we have 3 lots of 6 sevenths, we have 3 × 6 sevenths or $\tfrac{18}{7}$. Since each block of 7 sevenths makes up a unit, we write $2\tfrac{4}{7}$.

Although this explanation seems easy, we recommend that, as always, such results should be produced and illustrated by the use of diagrams and structured apparatus. What is clear given an understanding of the foundations is not clear if this is lacking. The instruction 'to multiply a fraction by a number multiply the top only' is not enough: it must be set against a confident knowledge of the roles played by top and bottom digits in the notation.

In constructing examples for class use, the essential denominators 2, 4, 8, and 16 can be extended with other one- or two-digit numbers since the principle is so simple, but there is still no good reason to discuss more than an occasional example of larger denominators.

The fraction as scale factor

The changing status of decimal and vulgar fractions not only calls for a reordering of the traditional syllabus, but suggests different emphasis within each topic. Pupils in the past, the numerate and innumerate adults of today, did the Four Rules in vulgar fractions during their elementary education, meeting other uses of fractions later and usually incidentally. The situation reminds us that arithmetic, although its study was stimulated by the needs of commerce after the Renaissance, was largely developed by scholars who looked on trade and technology as the business of shopkeepers and rude mechanicals.

We propose that before any work on multiplication and division of fractions begins the fraction should be considered as a scale factor. This is a more elaborate concept than either the simple comparison of two quantities or the 'parts of a whole' approaches so far discussed, yet it is the form in which one is most likely to meet a fraction in formal use. An educated adult should certainly be familiar with this use. It also happens to be, in this form, a useful starting point for the calculus, one of the branches of higher mathematics that the more able pupil may come to later in school life.

The fraction used as scale factor calls for a change in the meaning of a symbol. In spite of many equivalent forms, it is today sufficiently consistent to be regarded as standard in the context. It had its origin in the need for maps, diagrams, drawings, or models, made to a numerical scale, so that from measurements taken on the representation, the actual sizes or distances involved could easily be found. What is being compared is not the measure of a whole with a part or a larger with a smaller, but the measure of an object and that of its conventional representation as diagram or model. This standard symbol, when it is used instead of some other written equivalent, is the multiplication sign x *preceding* a number, an integer, decimal, or vulgar fraction. It is then usually read as 'by', and not, as before, as 'lots of'.

Thus, in a book which shows structures seen through a microscope, diagrams would be labelled x20 or x1000. This has a meaning, less obvious than it first appears, requiring discussion with children who meet it. It means, or should mean, that every linear measure is increased by the factor given when passing from the actual object to the picture, diagram, or model. Hence the true dimensions are 1 twentieth or 1 thousandth of those that appear thus magnified.

At one time, cheap toy microscopes were on sale which claimed to magnify one hundred or one thousand times, using a conveniently misleading alternative convention that the 'size' of an object was its projected *area*. We shall take it that the concept of a scale factor, the number that the dimensions of object are multiplied by to arrive at those of a representation of it, applies only to linear dimensions. That is, we compare the distance between two points on the object with the distance between the corresponding two points on the representation, the diagram, or the map.

The concept of scale should not really be difficult to a modern child who from infancy has played with dolls' houses, toy motor cars, and other scale models. What is important is that from the use of maps and models should emerge this abstract numerical concept, normally an integer if the diagram or model is larger than the original, or a fraction if it is smaller. The scale factor gives quantitative precision to the words larger and smaller.

The common equivalent terms and notations should be noted as the opportunity offers. Thus, the reduction factor of a map is never given as ×1/100,000 but either as the fraction itself, the so-called representative fraction, or as 1:100,000. The notation ×1 is then seen as equivalent to 'full size', taking its place between ×2 which means twice actual size and ×$\frac{1}{2}$ which is half size. One might see a diagram labelled 'scale $\frac{1}{2}$', 'half-scale', 'half-size'; these and any others are actual usages, and must be acknowledged and seen clearly as equivalent. Older maps with scales in mixed units were either approximations, as the 1:250,000 map which was sold as a 'quarter inch to the mile', or are being phased out like the one-inch ordnance survey.

Extension to scale factors such as ×$\frac{2}{3}$ or ×$\frac{3}{2}$ may be done, at the teacher's discretion, either with careful explanation or on the tacit assumption that the meaning is now evident. We have already made explicit on page 140 the relation $2 \times \frac{1}{3} = \frac{2}{3}$, so that it can readily be accepted that the technique of preparing a ×$\frac{2}{3}$ drawing is to take a measurement from the object, divide it into three parts and then take two of them. Similarly ×$\frac{3}{2}$ means that we take half of a measure and multiply this by 3. Note particularly the level of abstraction. We cannot take 3 halves of a given cake, but we can, when dealing with number, calculate 3 halves of a given measure.

Once more we can now bring in the fundamental mathematical concept of an inverse process. It is true that the explicit development of this concept, using the term itself, may go beyond the needs of the basic, but an informal approach is always worth attempting. If a drawn diagram is redrawn on a scale ×2, on what scale should the second diagram be redrawn to bring it back to the same size as the original? Such a question should be given with many examples until its meaning is quite clear.

A diagram will illustrate the process

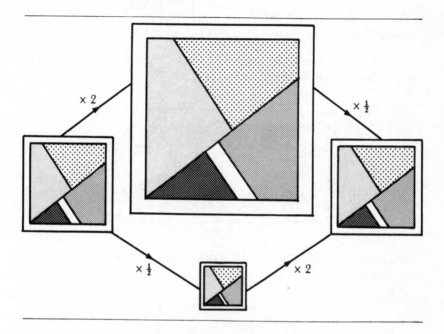

The scale operation ×2 doubles all dimensions, the operation ×½ reduces to half. If one is followed by the other, the diagram eventually returns to its original size. If, in the example, the side of the square frame were 3 centimetres, we should then have

3 × 2 × ½ = 3

3 × ½ × 2 = 3

Although the final result is the same, the intermediate step is different: this is at once seen from the diagram. We say that ×½ is the inverse scale factor to ×2, and ×2 the inverse to ×½. They are related exactly as multiplication and division. If we take 2 lots of 3 we get 6, and if we divide 6 by 2 we get 3.

We now have two ways of regarding 3 ×2. It is either thought of as 3 lots of 2 units, or as 3 units doubled. A useful and increasingly common teaching device is to think of the operation ×2 as that of a 'scaling machine'.

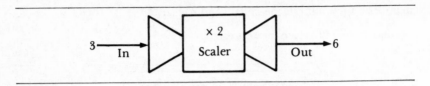

One feeds in the number 3, one gets out 6, and so on. This is a different representation from the earlier '3 lots of 2' applied to concrete objects

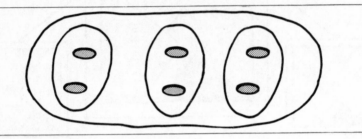

If the output 6 is fed into a second scaler whose factor is ×½, it emerges as the original input

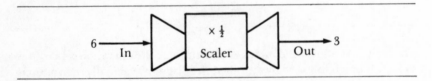

We can also think of a dividing machine

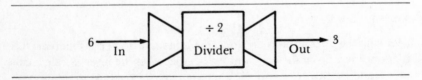

The labels ×½ and 2 are equivalent: they have exactly the same effect on the input. Linking together a ×2 machine and a ×½ machine gives a final output equal to the input, so that each of these two scales is the inverse of the other, as already described.

We have

$2 \times \frac{1}{2} = \frac{1}{2} \times 2 = 1$

and similarly for other scale inverses

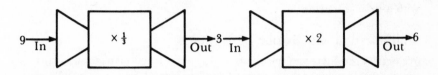

This is the two stage machine corresponding to the one below

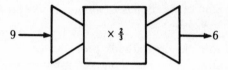

In the above example, 9 is reduced by the factor $\times \frac{1}{3}$ (or divided by 3 and then enlarged $\times 2$ (or multiplied by 2). The result is 6. If 6 is divided by 2 and multiplied by 3, we get back to 9

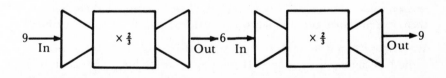

Hence $9 \times \frac{2}{3} \times \frac{3}{2} = 9$ and also $\frac{2}{3} \times \frac{3}{2} = 1$, so that $\frac{3}{2}$ is the inverse scale factor to $\times \frac{2}{3}$.

It may occur to readers that the foregoing treatment appears to be more abstract than is really necessary. This is proposed as a deliberate policy. We recommend that children who are unable to follow the presentation, made of course with ample examples and practical activities as given in any textbook of modern primary mathematics, should not be made to work with the multiplication and division of vulgar fractions in their general form. If the pupils have followed it, the next steps are greatly simplified, and in fact the learner is already over the conceptual hurdles. The next section begins with a brief discussion.

Multiplication and division of fractions

We regard products and quotients using vulgar fractions as topics which go beyond basic numeracy. In spite of their universal inclusion in syllabuses, they are, in fact, of very marginal utility in any but specialized topics in number work. Even at the level of teaching mathematics, as distinct from our present discussion of the content of numeracy, the processes could well be delayed.

It is easy enough to make up textbook problems which call for the multiplication of fractions; for example, one can ask for areas and volumes to be calculated for rectangles or cuboids dimensioned in fractions. But why? Although we are not here concerned with the needs of specific technologies or industries, we can ask in what actual technology are such calculations needed?

There is, nevertheless, the argument that a grasp of the process implies a number sense that may well be of value when insight as well as skill is called for. It is for this reason that we suggest teaching multiplication and division of fractions to pupils, basing the methods on the use of the scale factor. There is no doubt about the general utility of an understanding of scale, and we have suggested an extended treatment of this topic in an attempt to deepen understanding. If this attempt has been successful, then, but only then, can products and quotients of vulgar fractions be simply explained.

We also recommend that any examples used for teaching or practice should be numerically simple. If the object of teaching arithmetic is to provide material such as the example on page 90, for the benefit of people who like numerical doodling such examples have a function. But they are no more pertinent to an adult sense of number than crossword puzzles are to a sense of literary style.

We sketch in a possible classroom approach based on the use of the scale factor, although there are other methods with which teachers will be familiar. The product $3 \times \frac{1}{8}$ has already been seen in two ways, as 3 lots each of one-eighth of a unit, and as 3 units, specifically a *measure* of 3 units, scaled down by one-eighth. The second of these processes can now be applied to a measure which is itself fractional. A diagram helps to make the stages clear.

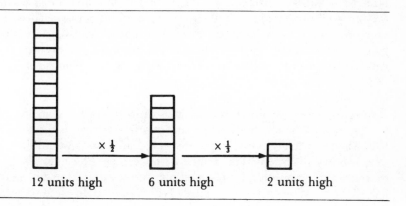

$12 \times \frac{1}{2} \times \frac{1}{3} = 2$

This could be done in one step as

$12 \times \frac{1}{6} = 2$

It follows that the product of $\frac{1}{2}$ and $\frac{1}{3}$ is $\frac{1}{6}$. A second diagram can be used to show the effect of commuting the terms, as on page 107.

The operation can be considered in four separate stages for numerators other than 1.

Thus

$\times \frac{3}{4} \times \frac{2}{5}$

can be interpreted as the four processes

1   Increase by scale factor ×3
2   Decrease by the factor ×$\frac{1}{4}$
3   Increase ×2
4   Decrease by $\frac{1}{5}$

This gives the overall change of scale ×$\frac{6}{20}$, as obtained directly by taking the products of numerators and denominators. This, of course, is what we all actually *do* when asked to multiply two fractions together. It is not suggested that all school-leavers should be able to explain the process in detail, only that they should at some time in their lives have followed the explanation, seeing that this is a process to be applied and not a trick to be remembered.

It will demand skilled teaching and carefully chosen examples to get this point home with understanding. A child who fails to follow the stages when they are well presented is not likely ever to need the process. There is no objection to the formulation of the rule once it is clear that this is what happens when obtaining the product of two fractions. We do deny that a child is being made numerate merely by being given the rule and drilled in its use.

If there are few general uses for products of fractions, there are even less for quotients. One might ask how many quarters or eighths there are in a half. This is formally equivalent to division, but nobody would do it this way, nor try to explain the result to whoever put the question by so expressing it. One would simply multiply, and say that 2 lots of $\frac{1}{4}$ make a half. If the question came from a puzzled child, one would use apparatus or a diagram.

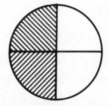

In mathematics one often needs to divide one ratio by another, but this application is hardly basic and is usually algebraic rather than numerical.

We shall nevertheless include simple division among the basics, as we did multiplication, because of the insight it gives. If a child is drilled in nothing but a rule which produces the required answers, he will get them, but only for as long as the questions continue to be set in the form in which he has learnt them. This is not numeracy.

Division is the inverse process to multiplication. To multiply by 3 and then divide by 3 leaves a quantity unchanged at the end of the double process. The pupil will already have met ×3 and ×$\frac{1}{3}$ as scale factors inverse to one another, and ×$\frac{1}{2}$ as the inverse to ×$\frac{2}{3}$ as on page 107. To multiply a quantity by $\frac{2}{3}$ and then divide it by $\frac{2}{3}$ leaves the original unchanged, and 'divide by $\frac{2}{3}$' is already established as ×$\frac{3}{2}$. If this is expressed as a rule we have:

To divide by a fraction, multiply by the inverse of the fraction.

This is, of course, equivalent in practice to the old 'turn upside down and multiply', which once again is what we all actually *do*. We form the inverse of a fraction by interchanging numerator and denominator. It is nonetheless true that the first form implies understanding, the second merely a trick for getting the result.

It is clear that this is a difficult conceptual step. It is neither easy to explain nor easy to follow if it is taken quickly. The teacher, who may of course prefer other approaches, will be able to adjust the pace to the abilities of the pupil, devising examples and illustrations of the process. If in the course of the work, the pupils' ideas of scale are deepened still further, an advance towards greater numeracy has been made. If, after all attempts, the pupil does not get it – and it is only too easily demonstrable that whatever teaching methods have been used in the past, many pupils did not – the duty of the school becomes clear. It is not to drag the pupil on through further stages of incomprehension by mechanical drilling, leaving the child to revert as soon as it ceases, but to concentrate on what we have argued is basic and essential. There could even be a second chance for the pupil later, since failure might well be the result of too hurried a progress through the vital first stages.

Summary

Those readers who have studied the flow diagrams used to summarize the work of each chapter will note from the one following that the interconnection of the various topics in dealing with fractions is now much more complex. There are more stages for the pupil to grasp before he has grasped the whole, and it is for this reason that the traditional order of fractions then decimals should be reversed.

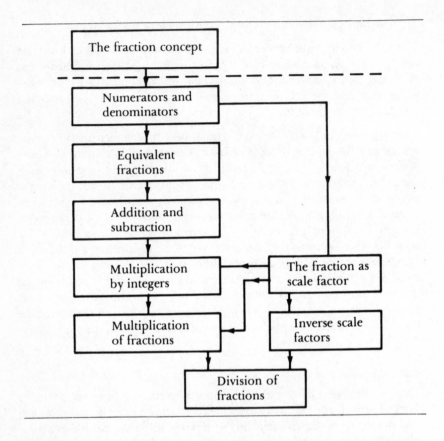

CHAPTER 8

NUMBER IN USE

Extended computation with decimals

All the calculations given so far have used numbers of a few digits, chosen to fit the stages reached in the discussion. The general skills required of school-leavers, and the applications to be found in everyday living and working, are likely to require little more. Although we do not regard it as part of the essential equipment of basic numeracy, we shall briefly discuss extended computation. By extended computation, we mean any calculations involving decimals more difficult than those already discussed. Here are a few:

1 0.036×0.279

2 4732×0.00027

3 $0.00126 \div 15.24$

4 $\dfrac{890 \times 0.963 \times 9.81 \times 3.14 \times 10^{-11}}{8 \times 1.17 \times 10^{-6} \times 0.82}$

Of these, the first three are examples from pure number, the fourth (which assumes a knowledge of the index notation) is an actual calculation that gives the viscosity of oil running through a capillary tube at a measured rate. How does one cope with these? The immediate answer, to which we devote a short section later in this chapter, is that if such calculations are needed in daily business the desk calculator performs them in the most economical way. This is not an adequate reply. The real problem is that the calculating machine has no number sense other than that provided by its operator. Only he can see, if he happens to press the wrong keys, that since the result for Example 1 is going to be a bit bigger than 0.006, 0.793 or whatever else may come up on the display *cannot* be the correct product. It is the 'feeling for number' that keeps computation under control.

Drilling at school, with rules about dealing with the decimal point repeated and remembered, may give a temporary skill in getting answers even to those who will never develop this feeling, but a break of a few years usually finds the school-leaver in difficulties.

Those who, in industry or commerce, have cause to compain about il-literacy and innumeracy, are not of course speaking with examples such as those given in mind. They are usually referring to school-leavers who can-not add two numbers together. Such school-leavers, however, may owe their plight to the lack of a programme devised for them based on an agreed scheme of what basic numeracy requires. They have, indeed, given up the attempt to follow such computations as those shown, and as a result have given up arithmetic altogether. If this is true, the majority of in-numerates are made rather than born.

Industry and commerce could, probably with advantage to themselves as well as to the community, provide their own training in specific number skills for their less able entrants. These will at least be motivated by the need to hold down a job.

The responsibility for the basic number knowledge remains with the schools. The less able will not acquire this knowledge by having their heads banged against what is for them the brick wall of extended calculation. A high standard of basic numeracy certainly calls for a careful reassessment of what schools actually try to do with their less able pupils.

Measure and units

Our traditional units grew up to meet the needs of individual trades and localities, although the spread of trade and manufacture, and the in-convenience of local variants of measures given the same name, led to early attempts at standardizing. Such examples as one can find in museums show that standard measures remained approximate in practice. It was not until the mid-nineteenth century that measures were defined sufficiently precise-ly to allow heavy manufacture to become precision engineering, and it was not before the middle of the twentieth that a fully coherent set of fundamental units sufficiently accurate for all foreseeable uses was finally agreed.

Any discussion of internationally standardized and codified units is only marginally relevant to numeracy. It is a fact that in the past we had to learn to convert into one another different units measuring the same quantity in different contexts. This often produced meaningless sets of figures in the process, as when one converted measures given to the nearest mile into measures apparently made to the nearest yard or foot.

Under the SI metricated system all equivalences are multiples of ten or denary submultiples. Once we had to think of 2 yards as 72 inches, or 2 shillings as 24 pence. Now we only have to think of 2 kilometres as 2000 metres, or £2 as 200 pence. Although the actual conversion is now very much simpler than before, this stage of understanding decimal equivalences is of the utmost importance to children, and they will need a great deal of concrete experience to grasp it.

For many other people familiarity with the limited range of Imperial units that happens to occur in their work has become habitual. After our own National Metrication Programme is substantially complete by 1980–1981, and, perhaps more important still, after the current Metric Act in the United States has taken effect, these uses will phase themselves slowly out. Such pockets of use as will certainly remain in everyday life will require neither precise conversion nor computation.

From our standpoint of examining numeracy, measure will now have two important features. The first is that in any conversion the nonzero digits do not alter.

In Imperial: 2 yards 1 foot = 7 feet

In Metric: 2 metres 1 decimetre = 21 decimetres

In actual usage, of course, one does not write 2m 1dm as such, but either 2.1 m or 21 dm, as with the draught of ships. A vessel drawing 60 dm will just touch bottom in 6 m of water, so that the conversion needed with fathoms and feet is no longer called for.

The second, and most important, point is that the significant figures are preserved. To say of two town centres that they are 12 miles apart does not mean that the distance between them is 21,120 yards. It is customary to give road distances to the nearest quarter mile: the conversion suggests an accuracy to the nearest 10 yards. But 12 kilometres becomes 12,000 metres, which by accepted convention means to the nearest thousand.

Much arithmetic has been done in the past in complete disregard of this technically important issue, not only by pupils then at school and now adult, but by teachers and others who wrote textbooks. Here are two or three examples. They come from the same 1965 commercial arithmetic as the computation on page 90. Although their units are no longer relevant to commercial practice, they are given to show the kind of seminumeracy that is bewitched by arrays of digits. The examples and others like them have fashioned the attitude to number shown by many adults who are now criticizing what is done in the schools.

We print both the examples and the answers as given in the book.

1   Convert 0.5986 ton to cwt., qr., lb.
Answer 11 cwt. 3 qr. 24.864 lb.

2   Express 3 cwt. 1 qr. 21 lb. as a decimal of a ton.
Answer 0.171875

3   A plantation produced 400,000 lb. of rubber at an average cost of 11d per lb. Find the total cost.
Answer £18,333 6s. 8d.

The examples were, of course, constructed when Imperial units were still in full use: this is not the point. Nor are we discussing the possible practical bearing of the calculations. The point is that these examples ignore completely the concept of significant figures in measured quantities. The final digits of the results given are quite meaningless, unrelated to the digits of the examples. Whatever such examples taught, it was not numeracy but a substitute for it, much as writing letters beginning 'We are in receipt of your esteemed favour . . .' was a commercial substitute for literacy. This matter will be of considerable importance in writing modern textbooks. Some textbooks in the past used measure as an excuse for arithmetic, instead of using arithmetic as a technique for handling measures. We must be careful not to do the same. If one wants a pupil to obtain the product of 12.3 and 6.7, this can be set as an exercise in pure number. To turn it into what looks like a practical situation by asking for the area of a rectangle 12.3 centimetres × 6.7 centimetres is to ignore the international decision to standardize on millimetres for detail measures. The rectangle should be given as 123 millimetres × 67 millimetres. The reading of x as 'by' is common usage and must be accepted.

The centimetre is a recommended measure for teaching purposes, and one can have rectangles such as 13 centimetres × 21 centimetres. But if more precise measures are needed, the unit is changed to the millimetre. Work with pure number follows whatever schemes the schools may devise, but work with measure must be in line with its use in the adult world of commercial or industrial practice. Basic numeracy calls for familiarity at an agreed level with both. The proper use of measure is not easy for schools to teach, because they have to devise and control situations in which measure is involved, rather than handle actual situations that require it. How do we teach children to choose a unit with which to express a measure? If the unit we use is too big, we shall either be hopelessly inaccurate or we shall be forced, in imagination or even in fact, to divide the unit into subunits. If it is too small, we have inconveniently large numbers and the risk of losing count. If we have actually to lay the units down, as in trying to measure a large room with a foot rule, the inaccuracy introduced by cumulative small overlaps or gaps may be quite large.

All this needs to be discussed and illustrated. Children should be asked to devise their own techniques of measurement and to compare them with the results of using other improvised or standard devices – rules, tapes, trundle wheels, paces, spans. It is much easier for the teacher today, because a set of carefully standardized units that everyone agrees on, with numerically convenient multiples and submultiples to meet the needs of larger or smaller measurements, is now replacing our native measures. The four pages of closely printed national and local units of measure once printed in the *Encyclopaedia Brittanica* are in effect replaced by seven standard fundamental units of which only four are needed in daily life. The local units may continue in use and will undoubtedly be learned in use by children, but they need not be formally taught. This is the answer to the odd emotional reaction that mention of chains, roods, acres, fathoms, and the like seems to evoke in people who when young learnt them in tables of measure.

One can now write the entire list of measures needed for general as opposed to special uses on a postcard. Nothing else need be used or need be taught and, within five years from now (1978), very little will be. Readers are asked to examine this list carefully and to ask themselves if anything they feel ought to be added is sufficiently general to be taught to everyone as a social requirement. Here is the list of required measures which should be known and understood:

Measures of Length:	kilometre (km)
	metre (m)
	decimetre (dm)
	centimetre (cm)
	millimetre (mm)

Measures of Mass:	metric tonne or megagram (Mg)
	kilogram (kg)
	gram (g)
	milligram (mg)
Measures of Fluid Capacity:	litre (l)
	millilitre (ml)
Measures of Time:	all common units as now used
Measure of Temperature:	degrees Celsius (°C)
Measures of Area:	square metre (m²)
	hectare (ha)
Measure of Volume:	cubic metre (m³)

To this list one could add, as an easy extension, square and cubic centimetres and millimetres, but they are not in household use. Not many people actually use cubic metres or hectares, but one reads daily of hectares under cultivation or cubic metres of soil removed in excavation, and one should know what these mean. One should also know that a litre of water weighs one kilogram and a cubic metre of water one tonne. The formal distinction between mass and weight is not normally needed, but should probably be taught in the hope of changing ordinary usage in the future.

Colleges of further education cater for any special needs of entrants into industry and commerce. We are concerned to define the minimum knowledge of measure that is needed for numeracy. It is a useful feature of the Standard International system of units that what we have described as the minimum is also adequate for all normal social needs.

Any reasonably educated person does, of course, acquire more than the minimum, and the SI system makes this easy. When one knows what a millimetre is, it is not difficult to guess what is meant by a millisecond. If such things can be learnt, they are worth knowing, but their context remains specialized.

We do not regard the residual use of Imperial units – calculation with them, that is, as distinct from serving a pint or cutting off a yard – as part of the general problem. Of course industrial spares in Imperial measures will be needed for some years, but this is no reason why all school-leavers should be expected to do complicated sums with inches and feet. As time goes on, the provision of equipment made to nonstandard measures will become increasingly a special requirement.

It might also be necessary for industrial training departments to review the way in which their apprentices handle number. Even if Imperial units are needed, it cannot be good engineering practice to dimension drawings in inches and fractions, and then to require the dimensions to be converted to decimals because the control settings for machine tools are so calibrated. Why not design to a decimal module in the first place?

In practice, decimalization took over in many fields where schoolwork remained obstinately traditional. Petrol has been sold in gallons and tenths since the earliest automatic pumps, and surveyors took heights in feet and tenths long before the present change to a metric ordnance survey. The failure of the schools, particularly primary schools, to do thorough basic work in decimals has contributed to the alarm with which some people regard the onset of metrication.

By and large we can look forward to practical measurement in the near future that makes far easier demands on number skill, if, but only if, the schools ensure that their leavers can handle decimals at the levels discussed in Chapter 6.

Practical measurement

The requirements of basic numeracy are simply expressed. A numerate person should be able to combine numbers of the sort likely to be met in daily life using such of the four processes as are appropriate, without numerical error.

The second skill that can reasonably be expected is the ability to use simple measures, to use a tape measure or a kitchen scale and get consistent results that agree accurately with those of others. The ability to use measures is clearly different from the ability to compute with them. It is a physical skill that can only be developed by practice, and is more than a mere knowledge of units. It is one thing to know what a milligram is, quite another to be able to use a laboratory precision balance. This last, however, like the use of micrometers or theodolites, is a special not a general skill. It is the ability to make measures about the house that we want everyone to have, so that curtains fit and recipes succeed. The numerate person should be able to operate on simple measures by the appropriate arithmetical processes without numerical error, but he should also be able to make the measurements.

Anyone who has to deal with measurement in any technical sense will know that the concept of an accurate measurement is quite complex. There is indeed a complete branch of mathematics devoted to error: to systematic, random, and cumulative errors in measurement, to rounding-off or cut-off errors in computation, to errors in the transmission and recording of results.

We are once again faced with a demarcation problem. The naïve faith in the validity of arithmetic that produced the examples on page 118 is not enough, but any formal discussion of errors and their management is too much. It is, as always, a matter of number sense, the end product of mature experience following enlightened teaching. The approach in class would be the collection of various measures made by pupils, with a discussion of why they do not always agree with one another. Most teachers now do this in introducing measuring apparatus to children, so that in the end pupils realize that *all* measures are made to some nearest given unit or subunit. Schools should provide an abundance of measuring apparatus and the school-leaver of today should be able to use similar equipment which he may find in the factory or workshop. Any work done should, however, make sure that measures of length, mass, and capacity take precedence.

Many pupils who do science, metalwork, and so on do in fact learn to use special measuring equipment such as micrometers to give very accurate results, but the range suggested for all pupils as a basic skill is quite modest: rules and tape measures for length, scales for mass, measures for liquid capacity. Over and above the skills of using such measures comes the very important ability, in today's society, of reading graduated scales of all kinds. A school will have pointer balances, and should certainly have a digital clock device with adjustable display. Children should now be taught the 24-hour system as well as the 12-hour clock, and one would include ability to read a digital clock and a 24-hour travel timetable as basic numeracy in action. As with calculation, we must ensure that the less able are not swamped by and discouraged by attempts to bring them up to some externally imposed standard of expertise.

Percentages and averages

We have already introduced the simple fraction as a ratio between two measures, one of which is taken as the unit, but we did not suggest that the word ratio should be used in class. A teacher might introduce it as a general concept, but we would not consider it as essential to general numeracy. Some other school subjects, such as quantitative studies in chemistry, depend very heavily on the use of ratio and proportion; it could be that ratio should be taught in a context such as chemistry as well as in the mathematics lesson. Such an arrangement would help both mathematics and science. But certainly the topics go beyond the minimum requirements for social well-being that we are trying to establish.

There is, however, one use of ratio that should be understood by everybody. This is the special ratio called percentage. The two standard calculations, the expression of one quantity as a percentage of another and the computation of a percentage of a given quantity, certainly occur more often in practice than most of the other 'commercial calculations' of the arithmetic books. We nevertheless regard them as specific. If anyone needs them occasionally, they soon learn how to do them even if they have forgotten what they were taught at school. If they need them frequently, they use ready reckoners or calculators.

But knowing that a percentage is the expression of parts per hundred is a different matter. This is essential. We are surrounded by information given in this way: anyone could write down a dozen examples from the newspapers. Naturally, one tries to teach the calculation of percentages, but one *must* teach the meaning of the word. Not many of us need to express the number of unemployed as a percentage of the adult population, but we should all know what is meant by the percentage statement as it is commonly given.

In schools there should be an adequate introduction to the concept long before the topic is proposed for computation. Pupils should, for example, plant a hundred seeds and note how many germinate, cut out a piece of graph paper having a hundred squares and colour in some of them, and so on. Where, in the past, the topic has been begun by stating rules, such as 'to express one quantity as a percentage of another, write as a fraction and multiply by a hundred over one', failure to grasp the process has led to failure to grasp the concept. This is not necessary, and the population as a whole should have, for numeracy, a clear picture of what is meant.

The topic has, like the conversion of units, attracted much invalid usage. We have the harmless circumlocution '$33\frac{1}{3}\%$' used for 'a third', but we also have percentages used when the numbers concerned do not justify it. It is statistically wrong to return 2 out of a count of 5 as 40%. The use of percent implies, or should imply, that the sample is large enough to be divided into hundredths. It makes sense to say that 6 out of every 100 of a large population are unemployed, but not 6 out of every 100 if there are only 15 to begin with. It may be convenient to do so, but any inference is likely to be false. This point is difficult to explain to many who consider themselves numerate, but it is really up to those who use and calculate percentages not to present invalid examples to those who, the bulk of the population, merely need to interpret the information.

The word 'average' has two distinct uses. It is used loosely in a statement such as 'the average man is not interested in politics', and with apparent numerical precision in a statement such as 'the average life of privately owned motor cars is eleven years'.

Here again there are well-known techniques for calculating averages, and a less elementary statistical treatment of average that discusses arithmetical means, modes, medians, deviation, and the like. All of these go beyond basic numeracy, although one might reasonably expect a numerate person to know how to set about calculating, say, the average height of a small group. For basic social numeracy, however, it is a knowledge of what is meant by 'average' that matters, and indeed most literate people know this because they are used to seeing the word in context. It is a fact that the 'average' or arithmetic mean, accurately computed by impeccable arithmetic, is one of the most widely misused and misinterpreted terms of statistics, and that no programme for basic numeracy is going to be of much help. It is the numerate and the articulate who misuse the concept most seriously, ignoring the protests of mathematicians. Nevertheless, the word is with us and we must see that children and students at least meet it in appropriate situations. Most children would, in fact, learn to calculate averages in these situations, but we must do our best to see that the less able retain at least the concept.

A note on calculators

The cheap pocket calculator has little connection with numeracy as such: its existence and rapidly expanding use hardly affects the requirements of number knowledge that we have discussed, any more than the typewriter absolves us from learning to write legibly. What it does do is take over the labour of routine computation. There are those who argue that routine computation trains the mind and even stiffens the morals, but no business or industry with a need for calculation could afford the labour charges. The calculator is needed as a matter of economics.

A calculator to cope with the first three examples on page 115 can be bought for a half-day's pay of an office junior, and another half day will buy a spare machine in case the first breaks down. A week's pay will buy a calculator that will accept simple programmes. For about a month's pay to an office junior or a little more, one can buy a calculator that prints out its results. A firm that needs routine calculations can no more employ people to do them on paper than it can employ calligraphers to write out its letters. This is a matter of business economics, not education. The issue of numeracy remains unaffected. An innumerate person would find a calculator as useless as an illiterate would a typewriter.

There is, of course, an ongoing discussion about the use of calculators in schools, about whether they should be permitted in examinations or encouraged for classwork, but their impact is on the facility with which pupils handle extended, not basic calculation. It is also true that the cheap calculator extends the range of normal numeracy. The customer in a small shop has long been in the habit of checking mentally the total for a few items: the housewife buying a week's food supply in a supermarket could not possibly do this. It is not unusual these days to see a shopper with a calculator, clocking up each purchase to check the total being spent. The calculator is merely auxiliary to the numeracy of the user, not a substitute for it.

It is true to say that in society today nobody need do, continuously, tedious arithmetical calculations. If he does, he is probably working inefficiently, and his output could be increased tenfold by a chainstore calculator costing less than a cheap watch. The implication for society and its schools is that we must try to separate basic and irreducible knowledge from what is best left to the machine. The time saved in schools curricula can well be devoted to the study of mathematics proper, but the schools, society itself, and the 'customers' for what the schools produce – the employers, that is – must decide what the basic and irreducible is going to be. The time has in fact come when those concerned with teaching must try to separate what is best left to the machine from number skills we can reasonably expect all citizens to possess. The next chapter contains, in a very tentative form, an attempt to give a list of what ought to be achieved.

CHAPTER 9

EDUCATIONAL OBJECTIVES IN NUMERACY

Everyone possesses some skill with numbers. These skills form a spectrum of ability from the socially inadequate to the extensive and confident. In this book we have tried to describe adequacy, and to make each stage clear by discussing not only the uses to which one needs to put the skill concerned, but the kind of classroom approach most likely to succeed with citizens when young. We are not, of course, suggesting that teaching should stop once the pupil has reached the level of adequacy we argue to be basic and essential, only that for the mathematically less able everything must be done in an attempt to attain it.

In earlier chapters we have listed skills to be developed by children in school, and be retained by adults if they are to take their place as numerate members of society. These skills were given in the form of what are usually called 'behavioural objectives', which in this context simply means that they state what a person should be able to *do*, as evidence of what he may claim to know or understand. An item on a syllabus is merely a topic that has to be studied or taught, a behavioural objective specifies a required result. Thus, 'addition of decimals' is a syllabus topic, while a relevant educational objective could be:

A person should be able to add a three-digit decimal to a three-digit decimal without error, the decimal point occurring anywhere in the numbers to be added.

Until these objectives are defined, the interpretation of any kind of syllabus is left to the teacher or examiner. We are convinced that a too wide interpretation of arithmetical syllabuses has led to a few skilled workers with number sense and a large tail of people lacking adequate number sense. There are many more adults innumerate than illiterate, many more than the known distribution of intelligence would suggest.

A basic syllabus is not much help. It lists topics that a school should cover, but what we need is an account of what people should be able to do. So far, there is no generally agreed list of such objectives for basic numeracy. What follows is offered for careful analysis and full discussion among those who are or feel themselves concerned. It is to be read as an attempt to start discussion rather than as a final list of agreed recommendations, although we do feel that not much can be omitted and that a good argument is needed before any new topic is added. Expressed as a teaching programme spread over the years of school life according to the ability of the pupil, this list is also a course of basic numeracy. The items given, however, are the end points of the programme, and are not class topics in a teaching order, which would need to include interim objectives such as the use of structured apparatus. The list could be extended indefinitely by cross-classification and subdivision, and might then suggest an elaborate programme of tests that ticks off for each child the growing list of accomplishment. This is not its purpose, and used in this way it could lead to a new fragmentation of arithmetic teaching, as indeed such lists have done in some countries where behavioural objectives are elaborated and overemphasized. Numeracy is an integrated set of mental skills and understanding: it must, as it were, keep all the balls in the air at the same time. But numeracy is not easily defined except by listing the skills that make it up. The list is offered primarily as a definition and only incidentally as a programme.

The items are given in four sections.

I Operations with integers

A numerate person should be able to:

1 Read and write the number words up to any limit.

2 Read and write the corresponding number symbols.

3 Count objects by putting them into one: one correspondence with the number words up to any required limit.

4 Record the numbers so counted on lists or tables.

5 Use the zero to record a null count.

6 Use zero correctly as a place holder in any number expressed in the denary notation.

7 Give verbally, by immediate recall and without the use of aids, the sum of any two single-digit numbers.

8 Count in decades beginning at any number of one digit.

9 Give verbally without hesitation the products of all number pairs up to 9 × 9.

10 Give verbally by immediate recall the differences and quotients that arise as inverses of the operations in objectives 7 and 9.

11 Identify the role of any digit in a multidigit denary number.

12 Add without error, using an algorithm, any two multidigit numbers.

13 Subtract any multidigit number from a larger number.

14 Multiply or divide any number by 10 and 100 mentally with a verbal description of the process.

15 Multiply correctly any number up to 3 digits by any number up to 2 digits. Any error should be detected and corrected when its existence is pointed out.

16 Divide a number of a few digits by one digit, with or without a remainder.

17 Identify and handle situations requiring the use of the four rules with integers, within the limits already stated.

II Operations with decimals and fractions

1 Identify and read examples of fractions and decimals as they occur in daily life.

2 Use simple fractions to describe situations.

3 Estimate simple fractions in suitable situations.

4 Decide, using illustration or any other method, which of two simple fractions is the larger.

5 Identify and form examples of equivalent fractions.

6 Explain the equivalence of vulgar fractions and decimals and describe verbally the notational difference.

7 Record fractional or decimal quantities as required.

8 State between which two consecutive whole numbers a decimal

lies, and identify the nearest.

9 Identify the role of each digit in a multidigit decimal.

10 Use decimals to describe given situations.

11 Multiply and divide any decimal by 10 and 100, describing the process verbally.

12 Find integers which are simple fractions of given numbers.

13 Multiply a simple fraction by an integer of one digit.

14 Multiply any decimal by an integer of one or two digits.

15 Divide any decimal by an integer of one digit to a given number of decimal places.

16 Round off any decimal to a given number of places.

17 Form the products of integers and fractions.

18 Recognize and handle any practical situation that requires any of the above operations.

III Money and measure

A numerate adult should be able to:

1 Handle the currency without difficulty in all practical situations.

2 Give change readily for purchases. To do this efficiently under retail trading conditions would need a few weeks of practice under instruction.

3 Use rules or tapes graduated in metres, and centimetres or millimetres as appropriate, to measure under domestic conditions, e.g., gardens, rooms, dress material.

4 Use graduated dial balances to weigh articles or to weigh out required quantities within the usual domestic limits.

5 Use equal arm balances with standard masses to the nearest 5 g.

6 Use measures of liquid capacity graduated in millilitres to the nearest 5 ml.

7 Find the number of metres of carpet of a given width required for a room.

8 Read both 12-hour clocks and 24-hour digital displays.

9 Interpret and use timetables for all forms of public transport.

10 Calculate and record elapsed time as on a job sheet.

11 Use calendars or diaries to record appointments, holidays, and

elapsed months, weeks, or days.

12 Read and record temperatures to the nearest degree.

13 Use and read a clinical thermometer after, but not before, demonstration and practice.

14 Read other graduated scales as on car dashboards, oven, or central-heating controls.

15 Be aware of the connection between right angles, degrees, and the complete turn, and be aware of the equivalence of 45° and half a right angle.

16 Use during the current transition period the common Imperial measures, without computation or conversion.

17 Recognize and handle situations with money and measures that call for computation within the limits set by Sections I and II.

IV Other numerical concepts

An adult should:

1 Be able to interpret a given percentage as parts per hundred, and be aware of simple fractional equivalents.

2 Be able to interpret the word 'average' when used correctly in context.

3 Be aware that the concepts of percentage and average are open to misuse, but not necessarily detect fallacious arguments based on them.

4 Be aware that all measures are subject to error and tolerance, stated or otherwise, but not necessarily be able to quantify these concepts.

5 Estimate the size of the result for multiplication of multidigit numbers.

Any list of required knowledge can be criticized both for what it omits and what it includes. We do not think that anyone who can cope with most of these detailed requirements can possibly be socially disadvantaged by innumeracy. What is more, given these as a starting point, anyone who finds himself with special arithmetical needs, as an apprentice or trainee, can acquire the necessary skills with the aid of demonstration or informed instruction. We have deliberately omitted reference to calculators but note that people who regard themselves as 'slow with figures' soon learn to cope with the Four Rules using a pocket calculator.

We also want to make it clear that many mathematical concepts that may appear in technical courses are not properly elementary. We are concerned only with operations with unsigned integers, fractions with small denominators, and decimals as discussed. The concepts of positive, negative, or irrational numbers require, if they have not in the past always had, a carefully structured approach that takes into account who is going to use them and for what purposes.

We are convinced that any further elementary processes needed can be built up without further conceptual difficulty from the foundations given, and it is in this sense that these are both basic and essential.

APPENDIX

NUMERACY AND THE TEACHING OF MATHEMATICS

The difference between a mathematician and one who habitually does arithmetic is like the former useful distinction between a chemist and an apothecary. We hope that we have made it quite clear that we are dealing with arithmetical skills and not with mathematical education as such.

One of the more satisfactory features of school curricula since the Second World War is the extension of the study of mathematics proper to all schools and all age groups. This extension, and the use of mathematical methods and formulations developed during the past century and a half, together often referred to as 'doing modern mathematics', have made the subject at once more exciting to teach and more rewarding to learn. But the new teaching in mathematics, with wider curricula in all schools, has meant that the time once spent in manipulative arithmetic has been much reduced. To the extent that most of that arithmetic is now of no practical importance, this is excellent; yet the useless arithmetic gone overboard seems to have taken with it the basic number skills that society not only needs, but needs more urgently.

Society also needs mathematicians and many more who, albeit not mathematicians, can use the language of mathematics, in much the same way it needs engineers, chemists, or linguists as specialists, and well-informed laymen, who can at least understand or work with them. What we need is a general education in number which can be extended to special interests and abilities as they arise.

But how do we determine a general education? We come back to the difficulties discussed earlier. Most schools have made up their minds about this. They begin by making all pupils do all the main line subjects available: English, mathematics, history, geography, one or two sciences, an art or craft, a language. Then, at 13+ or 14, they begin to select or to encourage interests.

Those who are bad at a subject can drop it. The colleges providing initial training for teachers are now beginning to increase their demands for entry qualifications, but it is a fact that in the ten years or so from 1965–1975 some 40 percent of the total entry either dropped mathematics in the 4th or 5th forms or failed it in the G.C.E. examination or its equivalent. Since those who dropped mathematics usually dropped physics and chemistry as well, they in effect dropped all quantitative studies, making no further effective use of either their mathematics or their elementary arithmetic. This last is crucial. It would be important in its effect on the population at large, but remember that we are speaking about persons who are now adult teachers, mainly in the primary or first school sector, and responsible for the numeracy of others.

This would be a more tolerable situation if the first years of schooling paid particular attention to the needs of basic numeracy as outlined. This certainly does not mean that the schools should remove mathematics from the curriculum and concentrate on 'good old-fashioned arithmetic'. It does mean that when pupils who are thinking of possible careers drop mathematics among other subjects they find uncongenial, they will at least have achieved basic numeracy. It may not be the function of schools to provide employers with a suitably trained labour force, but surely we cannot accept that a normal five year old should spend the next ten years of life in schools without being able to use words and numbers well enough to become an effective member of society.

We need mathematicians, technologists, scientists (even if not in the rather academic form that schools and colleges are tending to produce them) so urgently that we must give every pupil the chance to find these studies rewarding and hence to continue with them. We also need many more of our citizens, who are not professionally concerned in mathematics or its application, to have a greater understanding of the characteristic ways of thinking that govern all quantitative studies. And those of us who are interested in mathematics would like all educated people to realize, as the Greeks did, that mathematics represents a movement of the human mind and needs no more external justification than poetry or music.

But for historical reasons the teacher of mathematics is also the teacher of arithmetic, of applied number skill. The relevant argument now is who shall take the initiative in deciding what shall be taught. There is much discussion of a 'core syllabus' in basic subjects, which against all the traditions of English education would be imposed on the schools from above. It is also only too apparent that the English educational tradition of classroom autonomy has failed to do the job that society can reasonably expect of it. It is not going to be easy for the teacher. Teaching mathematics to those who have declared an interest is much more absorbing and, at almost any level in school, college, or university much less pedagogically demanding, than teaching the basic number skills to children not all of whom are receptive to them.

What is needed is an agreement both on the part of schools and the rest of society on what needs to be taught before all else. This does not mean, and we shall no doubt have to repeat continually that it does not mean, that no other mathematics can be done until basic numeracy is achieved. We believe that a programme of general rather than intensive training is likely to be most effective, using nonnumerical mathematics to leaven the basic number work.

But while a wide range of elementary noncomputational mathematical topics is available, reflecting the interests of the teacher and the response of the pupils, the basic number work needs to be an agreed, even a nationally agreed, programme. And we shall also need a kind of collective agreement and much resolution on the part of mathematics teachers to make sure that these skills are mastered. It is the teacher who in the end will be held responsible.

It would be easy to set external tests and embark on official assessment procedures, all too likely to fail in diagnosing the difficulties of individual children. Assessment needs, in the first place, to be teacher-controlled and part of the strategy of teaching the individual. We have already argued this in our report *Teaching Primary Mathematics: Strategy and Evaluation* (Harper and Row, 1977).

Moreover, the responsibility must be with every teacher. If a fourteen year old who is not subnormal cannot multiply two-digit integers, it may be comforting to blame the primary school he came from, but attaching the blame helps neither him nor society. What he needs is very skilled attention. But it is also true that the difficulty should have been recognized and diagnosed earlier. One of the saddest comments on education as an integrated social responsibility is the habit of many secondary schools, who give their intake tests to 'find out what they know'. It should be possible to organize a child's education so that the primary schools know what is to follow, and the secondary schools what has gone before. It is not all that difficult, surely, to keep records, or to discuss transfer. While we have such procedures which do in fact reveal blanks in the skills of older pupils, it will prove difficult to reject demands for imposed curricula and external tests.

We feel that the schools have to show what they propose to do. One approach would be to announce an agreed programme of number skill, identified as basic and essential but not allowed to expand beyond these limits, that should take precedence over all other arithmetical or mathematical work, although only in the sense that one car 'takes precedence' over another according to the Highway Code. All mathematics teaching will go on, but it should not be possible to drop this basic work at thirteen even if the pupil drops mathematics. Within the limits outlined in this book, it should not be difficult for the average child, nor impossible for the less able, at least to reach these levels between the ages of five and sixteen.

If some child does it by the age of nine, this should not be regarded as setting a norm for all children. The only critical age is the school-leaving age. Given universal compulsory education, no doubt some pupils will, when all has been done and tried, leave school innumerate, but at present our failures are too many. Our suggestion is that with these pupils too much has been attempted: in giving up the struggle with arithmetic they have surrendered numeracy. A teaching strategy for such slow learners to bring them to the critical level before leaving school needs to be worked out, and might well call for a reorganization of secondary timetables that schools and parents would not find it easy to accept. That is the challenge and it has not yet been met. Whether it should be left to teachers and the schools, or made a direct responsibility of education authorities or governing bodies, is a matter for discussion and possibly for dispute. At the moment there is still time for the primary and secondary schools to do it, but they have got to get together and tell us what they are going to do. If they do not, the climate of opinion may soon deprive them of the chance.

Index

Topics listed in the Table of Contents have no index entries for the chapters and sections devoted to them. Entries are given when such topics appear in other sections.